Adventures in

The Macroscope

To the Reader

As you read this book, it is very important that you read it as a story and not as a science book. It was our intention to create a fictional world based on atoms. Think of it as "science" fiction. You are not supposed to understand *why* the citizens of Atomville do the things they do just from reading this book. You are, however, supposed to enjoy it. Real world science *is* weaved throughout the story. You can go to the Atomville website (atomville.com) to explore and discover the real world science concepts used in the book. We hope that you love Niles and Livvie as much as we do and plan to bring you more of the adventures of Niles, Livvie and their friends in Atomville.

Jill and Cindy, May 2009

Adventures in Atomville

The Macroscope

Story by
Jill Linz and
Cindy Schwarz

Written by Jill Linz

Illustrated by Warren Gregory

Visit the website
http://atomville.com

Manufactured in the United States of America.

First printing 2009

ISBN 978-0-9722623-1-6

Thank you Page

The authors would like to thank Nancy Willard, Bee Gregory and George Laws. Nancy read many early versions and gave us invaluable advice about telling a story. Bee brought Al, the cat, to life and edited many versions of this book. George designed an amazing cover and put up with all of our requests for revisions. Without their help and encouragement, this book would still be sitting on a shelf.

To all of our other colleagues and students at Skidmore and Vassar, especially Mary Odekon, Shannon Stitzel and Jenny Magnes, thank you for reading the book, contributing story ideas and giving much needed moral support.

Special thanks to the team at Hyperspective Studios for their original character designs of Niles and Livvie as well as the town and logo design.

Of course, we also thank our families, George, JR, Norman, Michael and Bryan.

This project has received support from APS and AAPT New York State Sections and the Phoebe H. Beadle Science Fund at Vassar College.

About the Authors

Jill Linz received her master's degree in theoretical physics from RPI in 1991 and has been a member of the physics faculty at Skidmore College since 1992. Her own experiences have been non-traditional, as she entered the field of physics with a background in classical music. In the belief that physics education should begin as early as possible, she created the Physics Outreach Project in 2000. This project has brought physics to elementary age students and culminated with the production of two broadcast quality videos.

Cindy Schwarz is a Professor of Physics at Vassar College. She has a Ph.D. in experimental particle physics from Yale University. She has authored several books and one multimedia CD-ROM. Her first book, *A Tour of the Subatomic Zoo*, won the American Library Association Award for Outstanding Academic Book and is now out in a second edition. She also edited and published *Tales from the Subatomic Zoo*, a collection of short stories and poems written by her Vassar Students where subatomic particles are the characters. Her passion is making physics accessible and interesting to all.

CONTENTS

Main Characters

Al *– A big black cat living in the Outer World.*

Niles *– A nitrogen who is best friends with Livvie.*

Livvie *– An oxygen. She is a member of the band, The O3s and Nile's best friend.*

The Twins, Ollie and Odetta *– Both are oxygens and friends of Niles and Livvie.*

Ms. Na *– A sodium and teacher of GAB class.*

Ernie *– Nile's closest electron.*

Ella *– Livvie's closest electron.*

Ol' Louie *– A lead who is so big and cannot move easily that all the other atoms call him a tree.*

Royal Benzenes *– They are the ruling family in Atomville and are all carbons.*

Penelope *– Also known as The Wise Old Proton. Penelope is a proton.*

Lord Neon *– A neon and Keeper of the Sand.*

AL MEETS NILES

Everybody in the house where Al lived had long ago decided that he was just a nutty cat – cute and loveable maybe – but nutty just the same. He'd start running through the house, chasing after nothing as far as they could tell.

But Al knew that wasn't true. There were things buzzing around him, and he'd been trying and trying to catch one for such a long time. He scampered through the kitchen, down the hall, up the stairs and back down the stairs again, all the time flicking his tail and batting his paws. Then one day he caught one! Being a cat, now that he had it, he really didn't know what to do with it. So he thought he'd see if it tasted good. Now, just as he started to put it in his mouth he heard a panicky voice saying, "Hey – hold it Bub! Wait a minute! Don't eat me!"

Al stopped in mid-air. "Huh?" he said. "What's going on here? Who's that?"

"It's me! Here in your paws! I'm an atom!"

"So what's an atom? And who are you?"

"I'm Niles, and I live right here in Atomville – right here all around you."

"Wow, that's really cool," Al said, excitedly. "You mean you have a whole family? And friends? And a whole town? Right here?"

"Sure," Niles chuckled. "What'd you think? That I'm the only one here? Wrong! We're actually a pretty good-sized group, if I do say so myself. How about you? Are you the only one here? Does every one look like you?"

By this time Al had let Niles go and he was zipping up and down, in and out and all around Al's head. "Hold still! You're making me dizzy," Al cried.

"I can't," replied Niles. "My mom and dad say I have too much energy and I'm always bouncing off the walls."

"So," Niles said. "Are you the only one like you?"

Al had decided it was much easier to listen to Niles if he lay down and closed his eyes. So that's what he did. "Well, I'm the only cat. But I have some humans, four of them to be exact, two big ones and two small ones."

"What's a human? I never heard of that kind of atom. Or a cat either!"

"We're not atoms! We're animals! Humans and cats are both animals, but different kinds of animals. Cats rule the world," Al said rather smugly. "Humans are here to serve us. They cook for us, clean for us and generally do our bidding. But there are all kinds of animals. You got your dogs and birds and bugs."

"Animals?" Niles questioned. "Sounds like atoms to me. We have all kinds of atoms – you see, I'm a nitrogen, my best friend Livvie is an oxygen and our teacher is a sodium. We're all atoms, but we're all different. Cats seem a little like carbons if you ask me. "

Al thought about that for a minute. "You mean carbons rule your world?"

"Well, the King and Queen are carbons, but carbons in general don't rule – they only think they do. They think they're better than everyone else."

"You think I could come and visit you?"

Niles laughed out loud at that. "Naw, you're waaaay too big." He paused a bit, then he blurted out "Say, I could tell you all about it! The others are pretty nice and funny! Boy, could I tell you stories–"

"Ohhh, tell me how you got here!" Al mewed.

"OK!" Niles answered. "But it's a long story."

And so began an odd pairing – a large black cat named Al and his friend, the tiny orange atom named Niles – a friendship between two worlds – swapping stories, sharing laughs and, in general, doing what friends do.

NILES GOES TO SCHOOL

It began like any other day – I woke up in my room and had to get ready for school. Ugh – I had to go to GAB class that day. That stands for Good Atom Behavior. I always hated that class, constantly being told what we can do, what we should do and worst of all what we can't do. You see, atoms have certain rules and all young atoms have to learn to obey them. The rules can be really complicated, but – OOPS! That's not what we are supposed to talk about. Anyway, that day, like all days, the stupid hydros were flapping around singing outside my window. They're always building nests and singing outside my window.

My mom and dad made me my favorite red photons for breakfast that morning. There's nothing like a good meal of photons early in the day to give you energy for doing stuff later. After that it was off to school I went. I walked past the aluminum bushes, and, as always, climbed on the silicon rocks on my way to school. That's when I spied Nellie, and I had to take my secret route to school. Nellie and I go way back. Nellie's a nitrogen too, and she's out to get me. If she isn't trying to kiss me – yuck! – she's trying to trip me or steal one of my electrons or something. I just don't know about her.

So anyway, I had to take the route that took me past my mom and dad's store. They own a pet shop – mostly hydros and stuff for hydros – you know, cages and hydro toys. But sometimes they sell helis and liths – not very often though. Oops - hah hah, I forgot what I was talking about... Oh yeah! So anyway, I had to go past their shop to avoid Nellie. Well, that took longer

’cause I had to sneak around back and through the Town Hall to avoid being seen. Well, that almost made me late, so then I had to run.

Anyway, I get to school and meet up with Livvie and her two friends, the oxygen twins, Odetta and Ollie. Livvie is an oxygen, too. Her name is short for Olivia. I knew the three of them were in bonding mode ’cause they were ignoring me. Suddenly the bell rang and out of nowhere this angry photon, full of energy, comes heading right toward them. There was no time for them to get out of the way. The stupid photon flew right between Livvie and the twins, knocking Livvie on her butt.

"Ouch! You stupid photon," Livvie yelled.

"Come on, Livvie, let's go to class," I said, helping her up. Livvie got up, rubbing her butt 'cause it was sore from falling.

"You ok?" one of the twins asked.

"Yeah, I'm good. I hate those mean old photons," Livvie replied.

Of course we all agreed that the old mean ones are some of the best tasting too. Too bad they're so hard to catch. You can hardly see them until they ram into you. After that, Livvie and I went off to behavior class while the twins went off to Atomic History.

photons
food, energy
breaking bonds

GAB CLASS

Our teacher for behavior class was Ms. Na. She's a sodium, and of course, that makes her very yellow. Personally, I like my own orange coloring. Livvie and the twins, being oxygens, have beautiful shades of blue and violet. My pesky electron, Ernie, as usual, started trouble again – I know I shouldn't listen to him. But, man, he can talk me into almost anything and make it sound like a good idea.

All of a sudden I heard "Niles!" and thought…Oh, yeah, I'm in class, I better pay attention to the teacher or else I'll get in trouble. So I looked around – and everyone was laughing because I was caught daydreaming – again!

The teacher was standing over me with her hands on her hips and an angry look in her eye and said, "Now, Niles, since you are so smart that you don't have to pay attention, can you please tell the class how to keep your electrons in order?"

Always the class clown, I answered "In order of what?"

"Ah hemmmm! " The teacher was getting angry. You could always tell when Ms. Na was getting angry because her electron cloud turned yellow and her face was all puffy.

"What were we talking about?" I asked.

She told the class that we were discussing the proper behavior of our electrons and that a good atom knows how to control their electrons.

I figured I knew this one so I chimed in. "*Well*, a good atom knows that his electrons must be in cloud form to behave. If you let them, they will become gnats and try to make you do stuff you don't want."

At this point, Ms. Na was easing up a little. I really thought she was going to blow a photon right there in class. Man, that would have been cool. Instead, she asked, "And how do we keep them in cloud form?"

This was easy, so I said, "By minding our elders and not plopping on furniture – 'cause otherwise they bite." The whole class laughed at this one. There's not an atom in the class who doesn't know what it means to be bit by an electron. Even Ms. Na laughed at that one and added, "Well yes, plopping on furniture will cause them to bite. But what about when they try to cause trouble?"

The whole class said at the same time, "Ignore them!!!"

Ms. Na had gone back into teaching mode. "Yes – if you do not acknowledge them, they will revert back to cloud form. To become a good atom one must learn to ignore one's electrons."

That's when Livvie piped up and asked *the* question of all questions. "Excuse me, Ms. Na, I have a question. Why is it that our electrons can go from a normal cloud to an actual electron and back again?"

Ms Na had no clue. She said, "Well… I don't know… they just do…"

Livvie wouldn't give up. "Then what are they really?" she kept asking. "A cloud? Or a pesky gnat? What are they really and truly?"

I can get a little jealous of Livvie every now and then. She's so smart. I never would have even thought about what stupid electrons really were.

The teacher was frustrated and said, "Livvie, you are asking things no one has the answer to, but the royal scientists think that it depends on how you look at things." That's when the bell sounded and class ended.

As the class was leaving, I caught up to Livvie. "Hey Livvie, good going – you really got Ms Na. She didn't have a clue." I was laughing so hard I smacked my leg. "I bet she thought it was a real question."

"It *was* a real question. Haven't you ever wondered what electrons really are? We spend all this time in GAB class learning how to control them – but they never tell us what they really are – what if the pesky side is how they really are? What if we are forcing them to be clouds? That would be just – mean."

By this time, we were joined by the twins, who heard what Livvie had just said.

"What's all this about who someone really is?" Ollie asked.

"Did you hear about Mr. Carbona?" Odetta chimed in.

Sometimes it's really hard for me to figure out which twin is which. She continued to say, "It's all over school. They say that he is leaving the school to bond with one of the Royal Benzenes."

Soon after, Ms. Na came out of the classroom and told us to move along. She didn't want us standing in the hall gossiping. Then, of course, Ms. Na asked Livvie if she could stay after school that day to help her.

AN AFTERNOON WITHOUT LIVVIE

Since Livvie had to stay at school late that day to help out Ms. Na, I was on my own. And since I was daydreaming in class, I didn't learn my lesson about keeping my electrons in check.

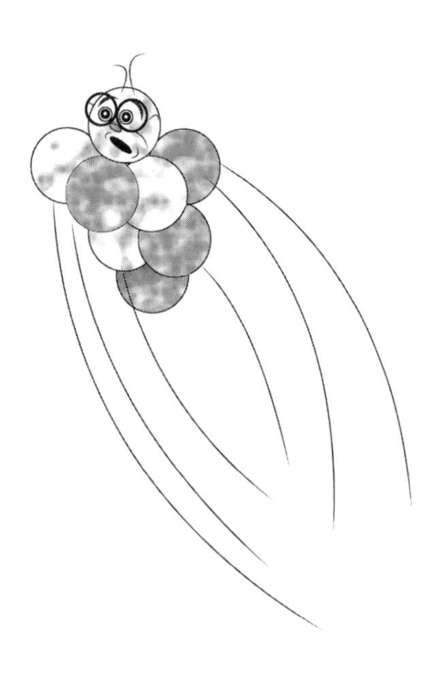

Ernie, my electron, who tends to get me in trouble, started

putting ideas into my head. "Don't ya wanna get even with that mean ol' Nellie? Remember at the dance last week when she tried to *kiss* you? You should go and mess up her room! Yeah, that's it – we'll mess up her room. That'll teach her not to kiss you, huh Niles? Whatta ya think? Huh? Huh?"

I said "Shhhh… I don't hear you…" and I began to sing "La la la la …"

Ernie started teasing me and kept saying, "That's not going to work... you know you wanna…"

I was still telling Ernie to shhhh – even as I walked to Nellie's house.

"Hmmmm… no one is home at Nellie's. I wonder if we can climb into Nellie's room?" Ernie said.

I told him, "We can't climb up – that would be breaking some law, I think."

"What law?" Ernie asked. Then he pointed to the nucleon trellis on the side of the house that led right up to her window. I walked up closer and looked up at the trellis. “Be careful to avoid the

sticky neutrons,” Ernie told me. The last thing I wanted was to get stuck by a neutron. That’s always trouble. The little red proton flowers are no problem, and they smell nice too. But those little green neutrons are something you want to stay away from. Trust me on this. Once, a nitrogen that I knew got stuck on one of those

with her friends, Odetta and Ollie, who are part of her band. Ooh, ooh, ooh – before I get to the real story I gotta tell you about music in Atomville. The most popular music is played by ozone bands. Ozone bands always have three players and they are always oxygens. The name of Livvie's band is the O3's. They are pretty popular as those bands go. Whenever they play, everyone in Atomville comes to see them. The O3's can play till all hours with everyone dancing the favorite Atomville dances. You have your symmetry dance, your asymmetric dance and your wags and rotation dances. Oh, and the wave! Atoms love to do the wave. The party can really get going when the O3's play. But something, or someone, always breaks the party up. Usually, it is some radical chlorine that comes in and causes trouble. They travel in packs – riding in on their bikes and scaring everyone away. It always seems to be the chlorine radicals that break up the party. Well, they really break up the band. It's probably just as well since the good citizens of Atomville need to tend to their everyday chores. If it weren't for the chlorine radicals, everyone would just dance, dance, dance. Oh, sorry, Al, back to the story.

Anyway, I jumped onto one of the bouncy oscillators and began bouncing up and down with the girls and Ollie. Oscillators are the favorite kind of ride in Atomville. Atoms love rides almost as much as they love dancing. Oscillators come in all shapes and sizes. The ones on the schoolyard are small ones, similar to swings, except they go up and down on a big spring. When the fair comes to town, the best oscillators do too. Bumper cars without any cars and the push me-pull me ride. There are even rides for the electrons – Jacob's Ladder is one of Ernie's favorites. He always makes me go on that with him.

The whole gang began to talk about their electrons. Livvie likes her closest electron, Ella, because she's the one that got her to form the band, and look how good that turned out. My closest electron, Ernie, is an OK sort. He does try to get me in trouble though, but I don't mind him. It's like he's a part of me. Everyone has a closest electron. Most atoms have a bunch of outer electrons. They come and go, like gnats buzzing around your head and then zipping off to do whatever it is electrons do, but your closest electron tends to stay with you all the time. It was all the talk about our electrons that caused us to question where we came from.

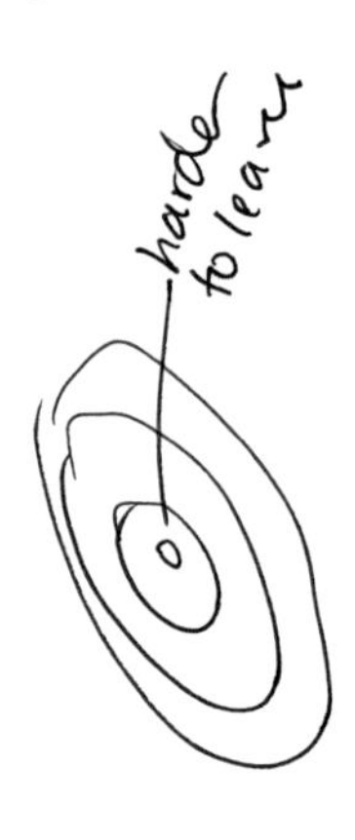

OL' LOUIE

Getting tired from all the bouncing, we all walked over to sit under the old lead tree. That old tree has been around for as long as anyone can remember. His trunk reaches up to the sky and his electron cloud billows out from the top and hangs down. He is a majestic old tree, but you should never, never try to touch his bark. It's well known in Atomville that if you get too close to the tree's trunk, that graceful looking electron cloud will come down and swat you away. And let me tell you – that hurts!

The old lead tree never talked much. Mostly he said things like "grummph" or "get outta hea".

Ol' Louie is what all the atoms call him. Mostly they just let him be. He is way too big to move around – so he's stuck right behind the oscillator park that is just behind the school.

As we sat on the ground, we began to talk about our families. Mine's pretty cool. My mom and dad are triple bonded. You know how strong that is. They do everything together. They say only nitrogens can triple bond. Not like oxygens. The best that they can do is double bond and that's only if they work for a molecule.

"Yeah, look at me," Livvie said. "I'm just about the only one I know who lives alone. Sometimes I wonder if I should keep a couple of hydros around. They really are pretty and they sing beautifully. I love to look at their bright magenta chest and their blue tipped wings."

I ignored Livvie's love of hydros.

Odetta piped in, "You're right, look at us, our whole family is made up of twins. There are lots of us in our house. We have six sets of twins plus our mom and dad."

Laying down, leaning my chin on my hands, I asked the next all-important question.

"How were we born?"

All three oxygens stared at me in awe. "What do you mean, how were we born? We just are. Life in Atomville has always been the same," one of the twins said.

The other twin piped up and said, "Niles, why do you and Livvie always come up with the strangest questions? First Livvie was all worried about her electron's true form and now you're worrying about where we came from. Well, I know where I'm going – home. You coming, Ollie?"

Ollie followed his twin sister. "Yeah, I'm hungry – and I think we're having blue-green photons for dinner."

"Yuck!" I told them. Those are gross. I'll eat any kind except blue-green ones. I had red this morning and I think Mom is making yellow photons for dinner. Now there's good eating. Just thinking about it, I started glowing redder and redder until suddenly a red photon burst out of my head, making a loud "burrrpppp" sound. "Ahhhhh," I said, "that feels better..."

"Ewwwwww...!" Livvie and Odetta yelled, while Ollie cracked up, slapping his leg. "That's perfect, Niles," Ollie said, still laughing – "blowing a photon in company! Ha, ha, ha!"

You see, Al, it's a boy atom thing. The grosser you can be the better. The girl atoms much prefer to be more polite about the whole emitting thing. Emission is a natural body function for atoms. Everyone does it. But polite company – that is girls and grown-ups – prefer to call it "emitting" Only gross boys call it "blowing a photon."

Still laughing, Ollie took off after Odetta, who thought it was a good time to get out of there. Left on our own, Livvie and I went back to talking when suddenly the old lead tree began to shake.

"Uh-oh – Ol' Louie's going to swat at us and we aren't anywhere near him," I told Livvie.

"Ah-hemmm…!" the tree thundered. "Let me tell you how I was born..."

"Huh?" Livvie and I said, spinning around to look at the tree. "Wow!" I said. "You never talk. Yell, yes. Growl, yes. But talk? Wow!"

He roared on, "I am the last of a great, great line of atoms. I know my family history all the way back to the beginning. My mother was a polonium, and her mother was a radon, and her mother was a radium, and her mother was a uranium. Each was a proud atom in her own right. The story was told to me by the Wise Old Proton."

"Who is that?" Livvie asked, blinking her big blue eyes in wonderment.

"The Wise Old Proton, she who has lived forever," the old tree shuddered. "She has all the answers. All you have to do is find her."

"How do we do that?" we asked.

"You must look closely to find her. The Wise Old Proton is very tiny. Sometimes it helps to use the magic magnifier," Louie told us. "Look through that and maybe she'll show up. I am tired now – go away!" Louie brushed us away with his electron cloud. But he didn't swat us.

"That was pretty cool," I said, hopping up and down as we walked home. "How are we going to get the magic magnifier?"

"We can't," Livvie said. "You know how the elders feel about such things. The only one I ever saw is locked up in Town Hall. Remember when our history teacher, Mr. Carbona, took us on that field trip to see it? They had it on display. It is said that you can see what we are made of by looking through it. The King and Queen thought that too much knowledge would lead to a revolution or something. So they locked it up and only the royal scientists can use it."

"Then the only way to get it would be to *steal* it," I said.

"I am not about to break any laws!" Livvie replied. Livvie always was the careful one. But even Ella, Livvie's electron thought it was a good idea. "You're not *stealing* it," she whispered to Livvie, "You're *borrowing* it. You'll bring it back.

No one will ever know, think about it – you and Niles can find the Wise Old Proton..."

Meanwhile, I jumped on the silicon rocks by the side of the road. Clearing silicon rocks is a never-ending job for the road crews because silicon rocks seem to appear out of nowhere overnight. So, the sides of the road are full of them piled high. This is a nuisance for many atoms but not for me. I love to climb on them and jump off over and over again.

THE BIG BORROW

We made plans to *borrow* the magic magnifier the next day. They kept it in Town Hall, in a closet. All we had to do was walk in and take it. Or at least that's what we thought.

When Livvie and I snuck into Town Hall to find the magic magnifier, we had no idea which closet they might be keeping it in. So, we started to look around. We went in the front door like we were just visiting. Town Hall has all kinds of rooms and things. There's the throne room for the Royal Benzenes to meet with us regular folks. That's the first place we looked. We weren't supposed to be in there but we looked around quickly and didn't see anything interesting. Then we went into the room where all the historical stuff is on display. We decided to start looking for the magic magnifier in there. We saw the atom family tree that shows that all atoms – no matter what kind of atom – are related in some way. On the wall were the big portraits of Mother Proton and Father Neutron – the parents of all atoms. And then, on the table in a glass case, were the Declarations of Independent States. Livvie and I started looking around at all the historical stuff pretending we were interested so no one would notice us.

"Livvie – Do you see it?" I whispered.

"No, shhh! Keep your eyes open. It has to be here somewhere," she whispered back.

Then we noticed a couple of mean old borons standing next to a little door. It was a tiny door – just big enough for a carbon to fit through. Carbons are smaller than me, but I thought I might be able to squeeze through if I could get past the borons. Borons are really stupid but also really mean and ugly and smelly! They have great big noses and snarly teeth and disgusting stuff

slobbering out of their mouths. That's why they make good guards. No one wants to go near them. They're also really hard to move. It seemed really weird that they were guarding that door.

"Do you think it's behind that door?" Livvie asked.

"It might be," I said. "Why else would they have borons guarding that door?"

"I'll distract them and you sneak in the door," Livvie said. "I won't be able to fit through, but you should be ok."

"Wait," I said. "What if the borons catch you and decide to keep you. What'll we do then?"

"You're right," Livvie answered. "What if they catch me? We gotta think of another way!"

It was Livvie's electron, Ella, who came up with the brilliant plan. You see, we can only talk to our own electrons, not other atom's electrons. So, like, I can talk to my electron Ernie, but I can't talk to Livvie's electron, Ella. Now the idea was that Ella and Ernie would try to convince the borons' electrons that the borons had to go and check out the throne room because they heard a noise in there. Ella and Ernie told our outer electrons to go join with the boron's electrons. Within a blink, we saw those little buggers' cloud shrink up to a little gnat and fly off to the boron cloud. Like I said, borons are really dumb. Their electron's can make them do anything. And, it was easy for ours to join in with them.

If any of our electrons leave, then somehow the Benzene's can tell where we are and we can get caught. I don't know why, but my parents call it being ionized. Anyway, the Royal Benzenes can see you when you are ionized so it's risky to let your electrons go off on their own. Livvie and I were willing to risk it, though, because it was the only way to trick the borons.

The mean old borons began to shuffle and waddle off to see what the sound was. As soon as they were out of sight, I ducked through the tiny door and into a tiny room. I couldn't stand up; all I could do is crawl along the floor. On the little table in the center of the room was a large pair of glasses. "That has to be it!" I said to myself. I also thought I saw something else in there whiz past me, but I didn't have time to find out what it was.

I grabbed the glasses and scurried out of the tiny doorway, slamming the door behind me.

THE MACROSCOPE

"Run!!" I yelled and Livvie and I ran as fast as we could out of Town Hall, down Main Street and then down Upp street and back to the playground to try the magic magnifier out. It looked just like a pair of glasses, but when Livvie put them on, her eyes looked really little to me. But to Livvie, everything was really close up. She tried looking around but it was making her sick to her stomach.

I wanted to try – "Let me! Let me," I said as I grabbed for the magnifier. Livvie pulled back so I wouldn't get them and as I grabbed for them, I knocked the magic magnifier off of Livvie and onto the ground. We both just stared at the ground and the broken magnifier. "What'll we do…? The lenses are busted out," Livvie cried.

"How are we going to fix it?" I exclaimed.

We tried putting the glass back in the frame – but unknowingly put the lenses in backwards. When Livvie tried it out she was astonished at what she saw! And so was I!

"Livvie! Your eyes are huge!" I said excitedly.

"Niles, where are you! I can't see you," she cried.

Livvie was walking around in circles waving her arms in front of her trying to find me.

"I'm right here Livvie," I said. But she didn't seem to hear me. "Livvie!" I yelled this time. At the same time, Livvie yelled too.

"Ahhhh – where am I?"

She jumped up in the air and started zipping around above me and yelling, "Something's after me! Help!" She pulled the glasses off and found herself safe and sound in good ol' Atomville. "Niles! You won't believe what I just saw!" Livvie exclaimed. She was all excited. "I had a big black furry thing chasing me."

"What was it?" I asked.

"I have no idea, but it was the biggest thing I ever saw!"

"Let me try," I said.

"OK, but be careful. You don't want to get caught by that thing," Livvie said.

So I put on the glasses and was amazed by what I saw. Luckily I didn't see the big furry monster, which, I might add, turned out to be you, Al. What I did see was a big kitchen. It looks something like our kitchen, but bigger. Waay bigger! I could fly all around in that kitchen. And the amazing thing was that I didn't see Livvie, or Atomville, or anything I was used to. Everything was giant-sized. I looked around for a while and then took off the glasses. Suddenly the world looked normal again.

I turned to Livvie and said, "This is sooo cool. I didn't see the black furry monster." Sorry Al, but that's how we thought of you until today.

"That's all right. Until today I thought you were lunch," smirked Al as he stretched his paws outward.

Anyway, Livvie and I were very excited about our new discovery.

"What do think we saw?" I asked Livvie.

"I'm not sure," She answered. "I have to think about this, but we better get this back to Town Hall before someone notices it missing."

"Hey, what do you think we should call this?" I wondered.

"I know," Livvie said. "Let's call it a MACROSCOPE!"

"A macroscope? What's that?" I asked.

"It's this! I just made that word up. But only you and me can know about this. OK, Niles? You have to pinky-swear never to tell anyone about this. Promise?" Livvie was verrrry serious, so I promised. And I never told anyone about it until now. But I guess you don't really count, Al. I don't think the royal scientists will be asking you about it too soon.

But the last thing we wanted was for the royal scientists to notice it was gone. Now, we hadn't thought about it, but when we *borrowed* the magic magnifier, we let a couple of our electrons go. We forgot that the royals could see us since we were missing electrons. Boy, we lucked out the first time. We had to make sure we had all our electrons before returning the magnifier. Electrons are everywhere in Atomville. Loose ones are always buzzing around, enough to drive you crazy. Livvie and I had to catch a couple. It's pretty easy, you just grab for them. Once you got 'em, they'll stay with you if you have room. Sometimes we like to catch them for the fun of it. Once, Livvie and I caught some and tried to sell them. We set up a little table by the side of the road and made up a sign that said "Electrons for sale – 10 cents." But no one bought any. So we were stuck with a bunch of electrons in a jar. We ended up letting them go 'cause they make squeaky, annoying noises when you try to make them do stuff they don't want to.

"So, what about the magnifier?" Al said.

Oh yeah, we caught a couple of electrons so we couldn't be detected and we headed back to Town Hall. I didn't want Livvie to get in trouble, so I went in by myself. The mean old borons were right back guarding the little room again. They looked like they were sleeping so I very quietly snuck past them and ducked into the little room. Well, something must have startled the borons 'cause they suddenly woke up. I was stuck in that little room with the snarly borons right outside! I was trapped! There was no way I was getting out of there without help. All I could do was to hope that Livvie came up with a plan to get me out.

THE WISE OLD PROTON

So there I was, trapped in the little room. There was a chair by the table, but it was too small for me, so I just sat on the floor. I still had the magic magnifier, or macroscope as Livvie and I called it, in my hands. I was fiddling with it while I wondered how I was going to get out. What if the royal scientists decided they wanted to use it today? Boy, would I be in for it then. There I was, just sitting there, when I thought I heard a faint "pssst..." At first I thought it was Livvie, but it didn't really sound like her. Then I heard it again, "pssst..." I looked around, but didn't see anything. I'm probably hearing things, I thought. Then I held the macroscope up to my eyes backwards. You know, so the ear thingies were sticking out. Not for any reason, just 'cause they were there. That's when I saw her. It was a proton. Not a full-fledged hydro. This one didn't have any electrons. It was kind of like looking at yourself without any fur.

"Without any fur?" Al suddenly opened his eyes. Then he shook his head. "Without any fur?" He repeated. "Yuck, I'd look almost…almost human!"

Anyway, I asked the proton, "Hey, what are you doing here?"

"I am being held captive by the Royal Benzenes," the proton replied.

"Held captive? But why? Are you working for the neons and argons? My mom and dad said that the Benzenes would lock up any atom who sides with the neons and argons. Why? I don't know, but that's what they said. I heard 'em."

"I am known by some as the Wise Old Proton. I like the *wise* part, but I'm not so sure about the *old* part," she said.

"The Wise Old Proton? We've been looking for you! That's how we ended up here! All this time, we were looking for the Wise Old Proton, and here you are – right with the magic magnifier!"

"Wait a second, who's we? I only see you," the Wise Old Proton replied.

"We, is me and Livvie, I mean Livvie and I. Livvie's an oxygen and is too big to fit in here, so she waited outside while I snuck in to put this back," I said as I held up the glasses.

"Ahhh, the magic magnifier!" she said. "How did you two find out about that?"

So I blurted "Ol' Louie told us about it. He said you told him all about where he came from and then me and Livvie wanted to know where we came from and so then we snuck in and *borrowed* it, but then we broke it and then we saw all kinds of weird stuff and then we had to get it back here and that's how I got stuck."

“Whoa, little guy,” she said. “ You’re talking so fast I can’t keep up. Tell me again, only this time slower and with a little more detail.” It was funny that this tiny thing was calling *me* ‘little guy’.

THE ESCAPE

So I told her the whole story. "Well, well. Now we're both stuck. Do you think your friend Livvie will try to get you out?" she asked.

"Oh yeah, we're best friends," I said. "I just know Livvie will get us out."

At that moment, we heard a light tap on the wall. "That's gotta be Livvie," I said. "Pssst, Livvie, is that you?"

"Niles? Can you hear me?"

"Yeah," I said as I put my ear up against the wall. "Can you get us out of here?"

"Who's us?" Livvie asked.

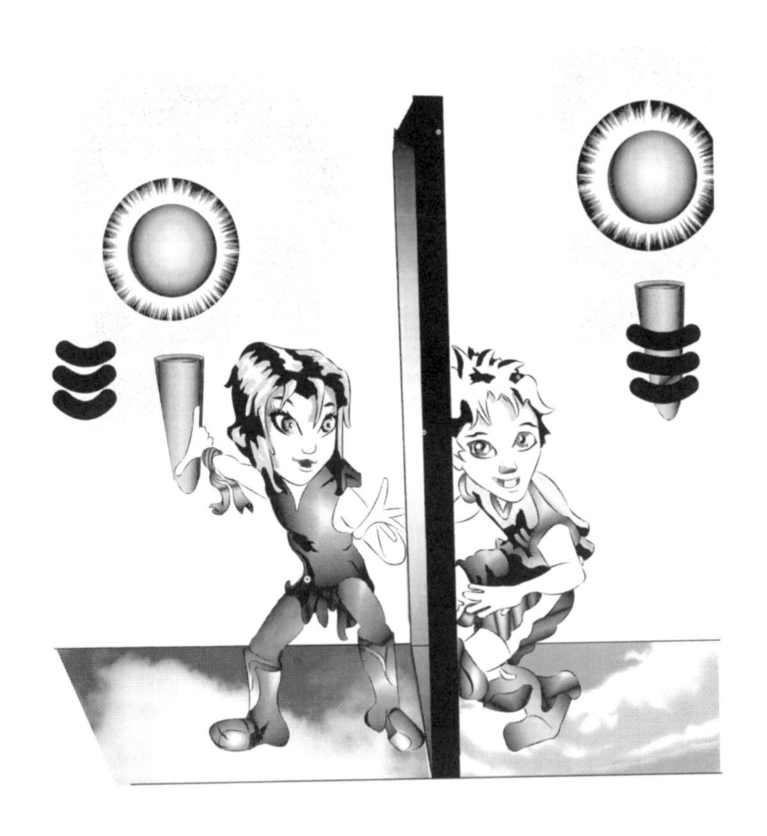

"The Wise Old Proton is in here with me! You gotta get us out."

"You're Kidding!? I will get you out," Livvie said. "Just give me a second to think."

"Wait," the Wise Old Proton said. "Didn't you say Livvie's an oxygen?"

"Yup, she sure is."

"I may have a plan that will get both of us out of here. Livvie won't be able to hear me through the wall, so, Niles, you will have to tell her for me. Ask her to come around to where the door is."

"I can do that," I said. "Livvie!" I called with my hands cupped around my mouth as I held them up to the wall.

Livvie replied, "What do you want me to do?"

"The Wise Old Proton told me that she and I were going to join forces and pretend to be an oxygen, I said. "We'll be a little unstable, but that's OK, we only need it to work for a short time. She wants you to go around and tease the borons. Then go and hide behind one of the pillars and stick your head out and tease them. Leave the rest up to us. When you see a wobbly oxygen, that's the cue to skee-daddle on out of there. You got that, Livvie?"

The Wise Old Proton perched herself on my head and I pulled my electron cloud up so it looked like we were one atom. The idea was that the dumb old borons would think we were an oxygen and confuse us with Livvie.

"Wait a minute here," Al interrupted. "Just how is that supposed to fool the borons? I don't get it!"

"Well, you see, in Atomville all nitrogens are sevens and all oxygens are eights. We are who we are because of how many protons we have. So if the Wise Old Proton pretends to be a part of me, the borons would see us as an eight – then they would think we were an oxygen."

Anyway, Al, then we opened the door just a crack so we could see out and sure enough, there was Livvie peeking out from around the pillar.

"Hey, borons!" she yelled. "Boy, are you guys dumb! I heard you were so dumb you need to be told how to eat!"

Now, those borons may be dumb, but they sure don't like being teased. They started snarling and making gross boron noises and began shuffling toward Livvie. That's when we made our move. The Wise Old Proton and I snuck out and ducked behind the other pillar. That's when we jumped out and the Wise Old Proton started calling the borons.

"Hey, borons... over here. What? Are you so dumb you didn't see me go over here?" The borons looked up at us.

"Huhhh?" they snarled.

Livvie took her cue and ducked back behind the pillar. The borons started shuffling over towards us. Then Livvie jumped out from behind her pillar.

"Hey, dumb borons," Livvie sang out teasingly.

Now those dumb old borons were *really* confused so that's when we made a run for it. We hightailed it out of there so fast, you might have thought we were struck by photons. We ran all the way back to Livvie's. We were all panting. Even Ernie and Ella were panting. I'll bet the borons are still wondering what happened.

NEW CLAYON

Livvie's room is really cool. She has a whole collection of music makers. She has more than any atom I know. She's got some string types and some blowing types. She's really good at the string kind. Sometimes I try to play with the little one, but I'm not very good at it. Livvie says she thinks the string kind is best. She calls them the har-monies and when she plays them, she says she can feel the whole universe talking to her. I don't know, Al, I can hardly make anything come out of them. And I never heard anything either. But Livvie can really make them sing.

When we got back to her place, Livvie started to play. She says it helps her think and calms her. The Wise Old Proton was so happy to see Livvie play that she started to dance around, waving her little proton arms.

"Oh Livvie, you can play!" The Wise Old Proton exclaimed. "Where did you learn?"

"I taught myself," Livvie replied. "A long time ago, I used to be bonded with an older oxygen. He gave me this one," she said, holding up her string music maker. "This one you have to play with a stick. The others don't use the stick. I kind of like the one with the stick the best."

When we finally had a chance to relax, we talked with the Wise Old Proton.

"Please don't call me that," she said. "I go by Penelope. The Royal Benzenes started calling me The Wise Old Proton when they captured me. I really come from New Clayon. There's lots of us there."

"New Clayon?" Livvie and I both asked at the same time. "We've never heard of that place."

"New Clayon is a place far from here where we live in peace and harmony. All atoms have shed their electron clouds and all electrons are free to live as they please – in gnat form," Penelope told us.

"You see?!" Livvie exclaimed. "I told you that electrons don't want to be in cloud form. I knew it, I just knew it."

"Why did the Benzenes capture you in the first place?" Livvie asked.

"Yeah, why did they capture you?" I added.

"Well, as you know, the Benzenes want to keep all atoms under their control. The way they do that is to keep all atoms from learning the truth. They keep everyone in the dark and make them think that the only way to survive is to work for the molecules. If all atoms knew they could do what they wanted, then the Benzenes would lose control."

Penelope then told us about New Clayon, where all citizens were free to do as they pleased. They all work together, gathering photons, cooking and cleaning and so on. The difference is that they can come and go as they please. When an atom works for one of the Benzene molecules, they are chained to it forever. She proceeded to tell us how she was in Atomville to speak with Lord Neon and the Duke of Argon who believe in a free and gaseous state – though they draw the line at losing their electron clouds. Anyway, she came to town to meet with them and form an alliance. They wanted to join forces to try and get rid of the Benzenes. On her way to the Duke's castle she was spotted by one of the Benzenes' spies, a carbon that goes by the name of Carla. Penelope didn't think anything of it at the time, but Carla detected her. You know how it is, no electrons and all. Well, apparently Carla ran and told the royal scientists because before she even got to the castle they scooped her up and placed her in that tiny cell where I found her.

"It seems like I was in there forever. It feels soooo good to be out. Now I have to figure out how to get back to New Clayon," Penelope said as she zipped around Livvie's room.

LIVVIE SAVES THE DAY

While we were trying to figure out how to get Penelope back to New Clayon, Livvie reminded us that Penelope could still be detected.

"You still don't have an electron," Livvie said. "At least for a little while, maybe you should take one on and pretend to be a hydro?"

"Yeah! I can run over to my parent's shop and get you a cage. No one will suspect if Livvie has a hydro!" I said.

"Wait," Livvie said. "Niles, you go and get a real hydro from your parents. If Penelope is just here by herself, even with an electron, I'll get all weird. If you get a real hydro, then the three of us can share electrons. Penelope won't have to pretend to have an electron and I can easily move around as a watery molecule. No one will ever suspect!"

"You are absolutely correct, Livvie," Penelope said. "We would be able to move around town with no one knowing. That's brilliant!"

So I ran over to my parent's shop and asked if I could have a hydro for Livvie.

"Livvie wants a hydro?" they asked.

"Yeah, she said she was getting lonely by herself and wanted some company," I replied.

"Wouldn't it make more sense if she just came and stayed with us? Most oxygens don't do well with just one hydro. Maybe she should take two?" my mom said.

"Uhhh, no, she doesn't want two. She just wants one hydro. Please, Mom and Dad? Please," I begged.

"Well, OK. But remember, if Livvie gets a little testy it will be because she only has one hydro."

So they gave me a cute little guy happily swinging on a perch in his cage. I ran with it as fast as I could back to Livvie's.

"Here ya go," I said. "Mom and dad are all worried about you, Livvie. They kept saying you should have two hydros. What'll we tell them when they see you with two?"

"They won't. We'll just have to stay out of their way until we can get Penelope back to her home town."

A NEW MACROSCOPE

So Livvie began to move around town as a watery molecule. Of course, we no longer had the macroscope and now we had to worry about getting Penelope back to New Clayon. Before

we did, though, we had to get in to see Lord Neon or the Duke of Argon so that Penelope could warn them about the Benzenes.

"That's all well and good," Al interrupted again. "But what about the macroscope? Remember? You were telling the story of how you invented the macroscope!"

"I'm getting to that." Finding the Wise Old Proton and the escape was all so exciting we completely forgot about the macroscope! It was good old Ernie who reminded me to ask Penelope about the macroscope.

"Penelope? Do you know anything about the macroscope...errr, I mean magic magnifier?" I asked.

She laughed at that and said, "Of course I do! I'm the one who brought the magic magnifier with me. I had it on me when they captured me. I was bringing it to Lord Neon and the Duke. I wanted to give them proof that we are made of smaller beings called the Quarks that live in the Inner World, and, using it backwards, as you say Niles, proves that we are smaller parts of an even bigger world! That's what you two saw, the Outer

World. That's the secret the Benzenes don't want the citizens of Atomville to know. Too much knowledge is a bad thing, they say. So they keep everyone in the dark and do vile things to those who threaten them."

"Wow!!" Livvie and I both said.

"But I guess it doesn't matter now," Livvie said. "We don't have it anymore."

"Yeah, we put it back. I left it in The Town Hall when we escaped."

Penelope started laughing at this too. "That's the beauty of the whole thing!" she said, clasping her tiny little hands together in joy.

"Anyone can make a magic magnifier! There's nothing magic about it. You have the power to create your own macroscope," she told us. "All you need is the silicon rocks that are strewn all over Atomville. If you feed the rocks with a lot of red photons all at once, the rocks will change into a magic magnifier. All you need to do is to catch a whole bunch of red photons. You know, the really dull, slow ones?"

Those things are pretty gross. They're really long and skinny and wiggly. They tend to bury themselves so you gotta dig them up. My mom and dad use them to catch hydros for the store. So anyway, that's what we did, and it worked! Penelope calls it 'heating'. I never heard of it, but it sure worked because as you can see, I'm here! We didn't make our own for a while, 'cause we promised Penelope we would get her to Lord Neon.

A VISIT WITH LORD NEON

The road to Lord Neon's castle was pretty long as the castle is located outside Atomville. Livvie and I had never left town before. My mom and dad told me there were gangs of atoms that roamed the outlands. They told me nitrogens and oxygens could get kidnapped or worse, so we were pretty scared to leave. As Livvie and I zipped down Main Street on our clouds, all the other atoms turned to look. It figured that town would be really crowded that afternoon. Normally we wouldn't have been noticed but Livvie looked so different. It's not that atoms had never seen a watery molecule before, but everyone knew Livvie

NEON CASTLE
OPEN

and couldn't believe she took on hydros. It was just so un-Livvie-like. So despite all our planning, we still didn't get out of town unnoticed.

"Let's move along," Penelope whispered. "I don't like everyone staring at us."

Livvie and I darted up Down Street and into a back alley. We went the rest of the way through town ducking in and out of alleys until we hit Strange Highway. That is the road that leads to Lord Neon's castle. I had no idea what to expect. I'd never seen a castle before but I had wondered what one looked like.

"Are there really gangs of atoms that will kidnap us?" I asked Penelope.

"Granted, there *are* bad atoms out there, but most atoms are kind hearted and mean no harm. The way I see it, there are worse gangs working for the Benzenes," she replied. "There are small groups of atoms that live in the outlands. They are generally very poor and don't have all the advantages that you two have. You don't know how lucky you are. Why, just gathering photons is a huge problem for them. You don't ever think about it, but not every atom can just grab photons and gobble them up."

Once we had gotten out of town, Penelope started flying around on her own. We let the hydro go 'cause we really didn't need it any more. I knew mom and dad would be mad, but I didn't want to think about that just then. For the most part, outside of town was wide open. We did see a few groups of atoms huddled by the side of the road though. They didn't have nice houses like us. There were little mounds that I guessed were houses. I had never seen such a thing. We saw a poor old oxygen couple that was barely moving. They just sat there outside their little mound. We zipped along that road for what seemed like forever. I was really getting tired and so was Livvie. It was starting to get dark as all the photons were beginning to bury themselves in for the night when we came upon a big group of lead trees, just like Ol' Louie.

"Well hello little ones!" one of the trees exclaimed. "Welcome to our little family. Won't you join us?"

"Why, thank you, Madam Leah," Penelope replied.

"Penelope!?" Madam Leah exclaimed.

"You know them?" Livvie said.

"Oh course," Penelope said. "Madam Leah and her family put me up for the night when I first came to Atomville. That was long before this last visit when I was captured."

"Captured?" Madam Leah said, sounding concerned. "Come, you must spend the night and tell me all about it. You two little ones can sleep in our clouds."

"Luke?" she yelled. "Can you form a comfortable bed for these two?"

Luke was another lead tree. He let his cloud flow down to the ground and form two very comfortable beds for us. We snuggled up in them and fell fast asleep. The next morning, we dug up a couple of photons for breakfast and continued on our way. I'll tell you what, Al, there sure were a lot of those silicon rocks out there and lots of hydros, too. We saw flocks of them flying around.

We began to see the castle in the distance. At first it looked really, really small. But as we got closer it got really, really big. It had a tower on it like the ones at Town Hall, but bigger, way bigger. The whole thing was made of silicon rocks but what made it so different was that it was made of the rocks themselves. Everything in Atomville is made from electron clouds. The houses, the furniture – even our clothes are all made from electron clouds, not from the atoms themselves. The castle got shinier and brighter the closer we got.

"Wow! Would you look at that!" Livvie cried. "I've never seen anything like it. Is this where Lord Neon lives? Did he build this?"

"This has been here for as long as I can remember. It is a very special place and Lord Neon has the honor of being the Keeper of the Sand. This place holds all the history of the atoms and the predictions of where we are going," Penelope told us. "It is here that we learned of the Inner World and the Outer World."

We entered the huge castle by walking over a bridge. There was a stream of red photons flowing around the castle. We went through the huge door and met Lord Neon. He was the reddest guy I ever saw. He was a big round guy and he wore a bright red uniform. And when he spoke, his voice boomed through the castle.

"Penelope!" He bellowed. "Where have you been? I got word that you were supposed to be here months ago." I kid you not, Al, the entire castle seemed to shake.

"I was captured by the Royal Benzenes and held in a room. These two helped me escape. Unfortunately, the Benzenes now have the magnifier I was going to bring you. We will have to construct a new one," Penelope told him.

Livvie and I got to catch the red photons to make the new magnifier. That was cool. We sat on the bridge and hung a line into the stream and caught them. Then Penelope took one of the silicon rocks. Not one from the castle though, a different one, and fed the rock the photons. Pretty soon that rock began to glow and then shrink. And when it was done, it looked like one of the lenses from the magnifier. Then we did the same thing to another rock. And pretty soon, we had a whole other magnifier, just like the one we had used.

"I knew it!" Lord Neon exclaimed rather loudly when he peered though the magnifier. "Now we have the proof that an Inner World exists."

"And don't forget the Outer World," Livvie blurted out.

"We think it's much cooler as a macroscope!" Livvie and I both said at the same time.

"We must get this to the Duke. With this device, we can convince the good citizens of Atomville that the true path is to live in a free and gaseous state!" Lord Neon exclaimed, holding the macroscope up in the air.

NILES AND LIVVIE'S SUPER SECRET MISSION

So much has happened since that first visit with Lord Neon. I'll have to tell you more later on. That's because Livvie and I are on a super secret mission to help Penelope and all the free and gaseous atoms. You see, Al, the Benzene's have been really bad. They have been kidnapping all kinds of atoms and making them work for the big giant molecules in the city. They also won't let any atoms know about the Inner World and the Outer World. Lord Neon is trying to become king so we can all live free and gaseous. So that's what we're doing, trying to find proof of the Outer World.

Now, just as Niles was explaining this to Al, two very large strange looking creatures came into the room.

"Those are my humans!" Al exclaimed. "I've got to tell them all about you, Niles."

Al suddenly got up and sauntered over to his humans. Really they were the younger humans – Sami and Jeff. Sami is a girl and younger than Jeff. Al jumped up on Sami's lap and began telling her all about Niles and Livvie.

"Meow… Meow… Rawrr… Rawrr," Al cried.

"What's the matter, Al?" Sami said stroking Al's back.

"Rawrr! Rawrr! Rawrr!" Al went on. He kept trying to tell them and all they heard was "rawrr, meow, rawrr." So Al jumped off and went back to the little rug he had been sitting on.

"That's one weird cat. It was like he was telling us a story," Jeff said, as Al walked back.

"So, what did they say, Al?" I asked.

"Ehhh, they don't understand my language," Al replied. "You see, that's why I'm smarter. I can understand them. I just can't speak their language. But they can't understand or speak my language. Very frustrating."

Just then, one of the humans, the girl I think, went over to a box and pushed a button. Then the two of them started doing the strangest thing. They started dancing around the room.

"Al! Look at that! They're dancing," I exclaimed. "If there is anything I know about, it's dancing. Remember I told you how atoms love to dance? But what are they dancing to? I don't hear any music."

"You don't?" Al said. "That stupid radio is so loud it hurts my ears."

"Hmmmm, I think I need to get Livvie. I'll be right back, Al," I said to him.

I took off the macroscope and was suddenly back in Atomville. I immediately went to find Livvie. I found her with Penelope.

"Livvie, Livvie," I cried. "You'll never guess what I saw with the macroscope."

"You used them by yourself?" Penelope asked. "We don't know what's out there. You shouldn't be going to the Outer World by yourself."

"Well, yeah, but you'll never guess what I saw. Livvie, you have to come with me. I met Al!" I said.

"Who's Al?" They both asked.

"He says he's a cat. I think it's some kind of atom. They have all kinds of atoms in the Outer World. I've seen them! Only Al says they call themselves animals, not atoms. And they dance! Just like us. I've seen them." I was so excited I kept going on and on.

"Hold on, what do you mean there are Outer-World atoms and they dance?" Livvie asked.

"You've got to come with me and see!" I grabbed a hold of Livvie and started pulling her. "Come on, I can show you. Al is waiting for us."

So we went back and got our macroscopes. By this time we both had one. We zipped on back to my house and on the way we saw roaming gangs of atoms. It looked like the Phonon gangs. They always cause trouble. Livvie and I think that is how the Benzenes kidnap atoms. The Phonon gangs come along and sweep up innocent atoms, who are never to be seen again. So, we ducked behind a big silicon rock until they passed.

"Do you think it's safe?" I asked Livvie.

"I don't know," She replied. "Just to be sure, let's go around the back of the Electric Boogaloo."

The Electric Boogaloo is a club for electrons. They can go there and dance. Electrons can't really dance by themselves. They have to go to a club to dance. When we passed by, we could see that they were having a really big party. All the electrons were in gnat form and dancing in a big line. They sort of have to as the club is real long and skinny. So all the electrons line up and dance back and forth. The dance they do consists of only a few steps. They move forward a couple of steps and then back a couple of steps. Seems a little boring to me, but then I like to zip around.

We passed on by the club and finally made it to my house and grabbed our macroscopes. We put them on and there in front of us was big and furry Al.

"Al!" I said. " I brought Livvie with me."

At first Al didn't respond. So I started zipping around his head. Suddenly Al saw us. He reached out with his paw to swat at us.

"Wait," I cried. "Al, remember me? It's Niles."

"You're real?" Al said rather surprised. "I thought I dreamed you."

"Of course we're real," I said. "Remember? I went to get Livvie?"

Al let out a big yawn and stretched. "Oh yeah, now I remember," he said.

Livvie was in awe. She had never, ever seen anything like Al before. So Al and I told her the whole story.

"Amazing," she said. "Where are the humans you were talking about?"

"There they are," Al told her. "Right over by the sink." They were still dancing to what Al calls human music.

"How come we can't hear it?" I asked Livvie.

"I have no idea, but maybe it's because we are so small compared to them," Livvie said. "I think we need to find a way to talk to them."

"Why?" Al asked. "I can tell you everything you need to know."

"I'm not so sure about that, Al," Livvie replied. "From what you said, I'll bet the humans can answer our questions. If only we could talk to them."

"Yeah," I chimed in. "Maybe we can find out where we came from."

"And, if our electrons really want to be in gnat form or in cloud form," Livvie added.

"And just what is the Outer World?" I said. "Can you answer any of those questions, Al?"

We were getting really excited thinking about finding out all kinds of stuff. But Al, we decided, really didn't know much at all. Just like a carbon. I decided that cats and carbons were really the same. By this time Al was getting dizzy watching us zip around and listening to our chatter.

"I am tired and bored. Now go away." He hissed as he stretched his back up in an arch.

We later learned that Al says he's bored whenever he doesn't understand.

"Do you think Penelope can figure out a way to talk to the humans?" I asked Livvie.

"I think it's worth a shot. Let's go back to Atomville and see if she can help us," she replied.

"Goodbye, Al. We'll be back. We have so much to learn about your world and much more to tell you about ours. Don't forget us *again*. We're not a dream. We're real!"

" Hey, Livvie," I said. "Lets get back to Atomville, Penelope needs us".

Livvie smiled at me, as best friends do and added, " And we need her too."

When the Sad One Comes to Stay

When the Sad One Comes to Stay

Florence Parry Heide

J. B. LIPPINCOTT COMPANY

Philadelphia and New York

Printed in the United States of America

First Edition

U.S. Library of Congress Cataloging in Publication Data

Heide, Florence Parry.
When the sad one comes to stay.

SUMMARY: Caught between her mother's fashionable world and her own deep-felt values, Sara must make a painful choice.
[1. Friendship—Fiction] I. Title.
PZ7.H36Wg [Fic] 75-9747
ISBN-0-397-31651-8

With very much love
to my son Chris

When the Sad One Comes to Stay

It is a warm summer day, and I am walking outside in our new neighborhood. I wear a pretty new dress, so if anyone sees me they will know I am going to be a nice neat neighbor, from a good family. Sally says things like that are very important. Sally has helped me do my hair a special way, and she has shown me how to tie my silk scarf and how to walk proud.

I don't know my way around this neighborhood yet, and I have no friends. Sally says school will be starting pretty soon, and then I'll have many new friends, nice friends, because this is a good neighborhood.

Sally is my mother. I have always called her Sally. She says if you call someone Mother or Mom or something like that, right away it changes things. "It creates a generation gap. If you call me Mother, I automatically play a mother role," she says. "That puts us both at a disadvantage. I want you to think of me as just one of your contemporaries, one of your peer group."

Sally's very clever that way. You can tell just by looking at her that she's intelligent. You can see that her brain is working all the time. You can feel it, whatever it is in her that makes her eyes dance and her body graceful and lively. She reaches out every second to touch more, see more, hear more, taste more. Leaning forward, jumping up, moving, plunging into actions, moods, forever trying something new, something different. Any newness is an adventure to her. I would like to be like Sally.

Sally has fixed up our new apartment so that it looks like Sally, bright and sharp and beautiful, and she has started to have new friends over, people like

Sally, handsome people, stylish people, excited people, people who are busy and plan things, people who get angry and argue. Their ideas and their excitements spill out into the room and fill it like smoke. After they leave, Sally paces.

"It's a whole new world, Sara, a whole new ball game," she says. "I'm on my way now. On my way! And do you want to know the secret of success, my pet? The secret of success is the sweet smell of success. It's knowing that it's out there somewhere, up there somewhere. Unless you know it's there waiting for you, you'll never see it, never."

Sally moves lightly, nervously, beautifully, around the room, emptying ashtrays, talking not to me but to the guests who are no longer there.

"Success," she says softly, testing the word, tasting it, letting its softness and its hardness seep into her, warming her and cooling her, strengthening her and weakening her, becoming part of her like her bones, the word itself and all of her now covered tightly by her beautiful skin. "Politics, Sara, a beautiful word, a diamond-sharp diamond-beautiful word, a word to conjure with. Power, my darling, power."

I am not sure what Sally does now, what her work is. It has something to do with politics. She represents a group of people, she is their spokesman, but what she stands for I do not know. Sally says I do not need to understand politics yet, I am too young.

Sally always sleeps late. She is a night person, she says. And she is often gone afternoons. This summer I am alone most of the time.

I walk around in the new neighborhood. I walk from our apartment building to the new empty school, and then I walk back. I try one way, then another, getting used to the streets, getting used to the neighborhood. It is not like anywhere we have lived before. Sally says this is a very safe, very respectable part of the city. The school is a fine one, very progressive, she has looked into it. Sally looks into everything. She tells me that I will make new friends, friends I can be proud of, friends who will give me advantages. Fine new friends who will open new doors. Sally seems to think life is a succession of doors to open, each door leads to still another corridor of doors, everything is ahead. There is no door of the past she wishes to explore. I try to picture Sally as a little girl, but I cannot. She never talks about her childhood, and I never ask her.

I wonder if any of my friends-to-be see me now from their apartment windows. I haven't seen anyone my age yet. Sally says maybe they are away at camp, or on a lovely trip. In a neighborhood like this, she tells me, people can afford to take a couple of weeks, maybe even the whole summer, and go somewhere—to the beach, maybe to the country or to the mountains. Maybe even abroad. "That will come for us if we play it right," she says. Sally talks as if everything is a kind of game—make the best move, play your cards carefully, throw the dice right, don't let your opponent see your hand.

The new apartment is very beautiful, my new bedroom is very beautiful. One of Sally's friends has picked out the colors and has chosen the furniture of my room. It is all pale blue and silver.

A room to be proud of, says Sally, but I have no one to show it to now, and I will be glad when I make friends. Sally is going to arrange for me to take ballet lessons this fall. I will meet nice girls that way. "The right kind of girls," she explains, "girls you can learn something from, girls who will do you some good. Contacts," says Sally. "Friends. People who count. Without them there's nothing. Remember that, Sara. It's not who you know, it's who you can get to know."

Sally's words are embroidered on my mind when I go to bed, like sayings woven into a sampler:

Use people or people will use you.

Let anyone need you, but never need anyone.

Don't sit back. Stand up and take, grab, get, get your share of what there is to get from life.

There's a lot of give and take in this life, Sara, so there are givers and there are takers. There are winners and there are losers. And I've never seen a giver who was a winner, Sara, never.

I walk around in our new neighborhood. We are in the same city as before, but it is a big city, and I feel that we have moved far away, to another state, to another part of the country.

By the time school starts this fall, I must know my way around. I practice now walking back and forth from our apartment building to the school, and I learn the names of the streets, and which streets run north and south, and which run east and west, and which streets have shops.

Between our apartment building and the school are some poor streets. There are ancient brownstones with loose bricks and steps you can sit on. I remember sitting long ago on steps like these to fasten my roller skates, to watch games of hopscotch, to play jacks. Sometimes, I remember, I would sit my doll on the steps to see, to help me watch, to keep me company. We have moved often since then, there was a city time and long ago a country time, but other days are not today and Sally has told me that old times are times no more.

The street I walk on now has a familiar feel. I am more comfortable here, but Sally will want me to walk to school the other way.

I hear a tapping on a window. I look up. An old woman knocks on the first-floor window of one of the brownstones and waves at me. I wave back and smile. She is setting something out on the window-sill, and I see that it is a plant with small green leaves.

I walk that way each day after this, and each day I see her. Each day she is doing something at the window, putting a plant in the sun or hanging something to dry. One day she motions for me to wait there. She signals for me not to go away. In a moment she opens the door, and her big loose comfortable body leans against the door frame. I stand on the first step and look up at her. She wears slippers and a big flowered apron. Her round face is wrinkled, her eyes are large and soft, like the eyes of my father.

"Is it you, Corrie?" she asks.

I shake my head. "My name is Sara," I say. "My name is Sara Goodhall."

She peers down at me. "How come you lookin' for me lessen you be Corrie?" she asks.

"I wasn't looking for anyone special," I say. I turn to go.

"You got the look of Corrie," she says. "Maybe you be Corrie, maybe just a different name."

I shake my head again.

"I already made Corrie some cookies," she says. "They be wastin' 'less you come tastin'."

I nod and smile, and I climb the steps that are like the steps I remember, and I follow her into the hall and into her apartment.

Her place seems like the place where we once lived, like the place of my father. There are soft frayed chairs and a lamp that could be the lamp we used to have, it is so like. The table in the middle of the front room is set for two.

She brings in cookies and hot chocolate from the kitchen, and then she sits on a rocking chair and asks me to sit at the table. She rocks and knits, and we talk.

I haven't had anyone to talk to for a long time, anyone but Sally and her friends, and that isn't like talking. With them I am always afraid I might say the wrong thing, might embarrass Sally, might amuse her friends. Always with Sally and her friends I feel that I am on a stage trying to remember a part I am supposed to play.

Now I talk. I say that we have just moved here,

my mother and I. I tell her we live in the Stanton Arms. I want her to know right away where we live, that it's a nice building on a nice street, one of the best neighborhoods in the city. I want everyone to know that. Otherwise, how can they tell that I am someone they will want to know, someone worthwhile?

She asks if my mother works, and I tell her that my mother is in politics, and that she is also interested in art and education and committees. I say that she is interested in the future, in what's going to be happening.

I keep talking, she keeps knitting. I tell her about the new school I will be going to, I tell her about the new dresses and coats and shoes that Sally has bought for me. She does not ask me about my father, and I don't tell her, I tell her only new things today. I do not think of telling of old times because I am not used to remembering them, and I have never talked of them. Sally does not like me to remember.

That first day I do most of the talking. She listens easy, like my father. Her name is Maisie Best. She lives here alone, but soon Corrie will be coming to stay.

When I get up to go, I thank her for the cookies and chocolate, I put my chair in place again, I do everything the way Sally has taught me. I am careful to talk the way Sally talks, and not the way my father does. I haven't seen my father in a long time, but I still remember the way he talks, and I remember what Sally has said about that, and about everything.

"Goodbye, Mrs. Best," I say. Although Sally wants me to call her Sally, she likes for me to call everyone else by their last name, to show respect, she says.

"Corrie call me Maisie," says the old woman. "It would be pleasable if you'd call me that."

"All right, Maisie," I say.

"Next time, you don't need to talk that throw-away talk," she tells me. "Ain' no need in talkin' 'bout just things, howsomever purty they be, or new. You can talk next time 'bout main things, things you keep sewed inside yourself."

Her smile is as contagious as a yawn. We keep smiling at each other as if we have a joke.

Every day now on my walks I stop at Maisie's. Every day we sit and talk, she knits, and sometimes we play cards, or checkers, or dominoes. It is easy to talk, the words fall soft and easy from me, and as I talk the memories begin, and I find myself wondering which came first, the words or the remembering. No matter, they are all one and the same now, and I talk faster and faster to keep pace with the pictures from the past that overtake me.

I tell her about the time of the rain.

It has been a long, hot and dry, dusty time, the front steps and the sidewalk too hot for my feet, and in the middle of the night the rain comes. I get awake on my couch and run in to where my father sleeps. We go outside in our nightgowns, in our bare feet. We stand on the steps, we run down to the sidewalk, to the street. We are the only ones outside.

"Our own private street," he says. "I bet you ain' never seen it so lonesomelike and beautiful before!" The parked cars glisten in the dark like huge silent

patient beetles. We run, laughing, our bodies tasting the rain.

"The whole street's havin' a free shower bath!" he says. He takes my hands in his, we dance in the rain that night, Puppie and me.

It is the first time I have said out loud to anyone the name I always called my father, and now I say it and say it again softly, *Puppie, Puppie.*

Maisie rocks and smiles and says, "Say a name soft and he come close by, say a name hard and he sure to die." She says, "My grandy told me that. You talk mean 'bout somebody, it sure to eat them up just a bit at a time, on the insides, not so's they notice right off, but so's they get sickly. And it works the other way 'round. You talk good, talk soft 'bout somebody, they feel nice. No matter how far off they be or how sad, all of a sudden they say, 'Hey, I feel good!' "

I smile at Maisie's stories. We play games, she tells me old memories, old jokes, we laugh together until our sides hurt.

"If the Sad One come by, see me now, his eyes would pop!" she chuckles, wiping her eyes.

She has not told me yet about the Sad One, that will come later. It is about the Sad One that I tell this story.

I ask her about her knitting. I think maybe I would like to learn.

"A present for my little granbaby," she explains. "A pair of booties for my little Corrie."

But I remember that Maisie had thought maybe I was Corrie when first she saw me, and I say this to Maisie now. "Corrie's not a little baby, is she?" I say.

"Sometimes I see her a baby," she answers, "but she's growed by now. Someday she be old like me. Not yet, but there come a sometime day."

"But she's not a baby," I insist.

"Baby, old lady, someplace in between, she the same Corrie, she don't change. She 'bout the same wherever she may be on the road. Where it don't count, she different, sure, she must be older, got to be bigger, but inside she still the same Corrie, like I tell you."

She looks over at me. "I think prob'ly she be 'bout your size, prob'ly she lookin' 'bout like you now."

"How long has it been since you've seen her?" I ask.

Maisie stops rocking, she stops knitting, she frowns. "A bit of time, a hunk of time," she says after a while. "But it go down fast, it go down fast lessen you look at it go. My grandy told me, if you start watchin' time then it start watchin' you."

She nods to herself. "Corrie may be older now, may be bigger, but she still my baby Corrie." She taps her head. "In here I can keep things just the way I like, throw out the bad stuff, change 'round what's left any which way I please."

Now she talks each day about Corrie. "The baby of my son, that's her. The baby of my baby. Maisie is the first word she ever say. I name her, she name me."

"But where is she now?" I ask.

Maisie shakes her head. "She be comin'," she says. "Just the way you come, she be comin', and I be ready."

We are easy together. Sometimes it is Maisie who does most of the talking. I listen to her the way I would listen to my father, soft words like cotton balls all coming together to make a resting cloud to lean on, to float on, and I stretch and sigh with contentment.

Today she tells me about Corrie's mother. "My son married one of them newfangled girls. Newfangled clothes, newfangled ways, newfangled notions. A pretty girl, a city girl, a girl with prideful ways."

She stops rocking. Her face rearranges itself, falls into sadness gently like a quilt sliding off a bed. I can see a sad face does not rest easy on her, and in a minute she makes her face comfortable again, she smiles.

"That Corrie. I'm surely lookin' for her now," she says. "Tomorrow or soon."

"How long has it been since you saw her?" I ask again.

But Maisie doesn't say. I think of my father and wonder if he waits for me as Maisie waits for Corrie.

Another day she says, "I be makin' her somethin' purty, a party dress. I make her lots of things." She gets up and goes over to a closet in the room, and I follow her. It is filled with dresses: baby dresses, little girl dresses, grown-up dresses.

"See, I make them a wee bigger every time," says Maisie, " 'cause she is bound to grow a wee bigger each day. That way I have somethin' nice for her when she come, no matter how growed she be. And I make the little ones too, for old times, 'cause that little Corrie be still inside the big one."

Every day the table in the front room is set for two. I say, "Do you set the table for me, Maisie?"

She laughs and rocks. "Maybe for you, maybe for Corrie. It don't matter none, long as it's set. Then if the Sad One poke his head in the door or peek in the window, he'll see I'm expectin' someone for company and he won't hang 'round none."

That is all she tells me this day about the Sad One, but she tells me a little every day after that.

"Maybe when you just a little tyke, maybe later when you middlin', maybe the Sad One pay you a visit. Maybe he just say how-do, size you up, but he don't linger none. He just come a little bit to get acquaintful. He don't stay permanent then. He only move in when you too old or sick or tired or sad to run." She laughs and rocks. "A body got to be foxy to keep him away then."

I like her stories. My father used to tell me stories. Sally never does.

"Can you see him? The Sad One, can you see him?" I ask.

"Oh, no, you can't see him," says Maisie. "Ain' nobody yet ever seen him."

"How do you know when he comes, then?" I ask.

She considers. "Well, you can sort of taste him. That's it, you can sort of taste him in the room."

She stops rocking. "I figured out ways to joke him," she says. She points to the table. "That's one trick I learn—settin' out two places. I know plenty others." She chuckles to herself.

"Tell me," I say.

She nods and leans forward. "I be tellin', but not all at one time," she whispers. "You got to slip it in sideways, sneak it in a little bit at a time. He sure to hear if you tell it all in a bunch. Then he know the way you fool him, and you can't never fool him again."

She teaches me to knit, and I knit many small squares of bright colors.

"We can sew all the little ones together and make a nice throw for your sofa, cover up the worn-out spots," she says.

I think of our elegant apartment and wonder what Sally would think of a patchwork cover like the one I'm making.

When I walk back today, I go the long way around. I pass apartment buildings as fine as ours. Maybe someone is looking out, maybe wondering if we can be friends. Maybe she is coming down now in the elevator, maybe she will call to me, run across the street to me, introduce herself. I practice my smile, I practice the way I will hold my head, what I will say. I have watched Sally when she meets someone new, and as I walk I try to think of things to say to make the new person want to know me.

My new friend will not want to hear about the time of the rain, and I will not tell her. I will tell her about the ballet lessons this fall, and ask if she will take ballet, too. I will talk to her about the new school.

No one comes, no one calls to me today, but I know soon I will be meeting my new friend.

I take the elevator up to our apartment. Sally is reading the paper quickly, nervously, her beautiful eyes moving restlessly over it until she has possessed it. She draws it into her, and when she has pulled from it all she needs, she casts it away, a useless shell.

Sally has breakfast, I have lunch, that is the way it usually is. Since Sally sleeps so late, she does not know how early I go out for my walks, so she feels that she leaves me only for the afternoons. She talks again about ballet lessons.

"You'll learn how to use your body. Control. It's the most perfect form of the dance."

I think of the ballet slippers that Sally will buy for me, and then I think of Bigun.

Bigun dances on the porch, the porch to our old house, the house of long ago, the country time before there was any city time. He dances, the dry wooden boards creaking under his bare feet. He claps and then I clap in time to his dance.

"Joy, Lillun, that's joy bubblin' up in the inside of us! See that tree, all them branches stickin' out? It's the

joy in that tree reachin' out, otherwise that tree'd be growin' like a telephone pole!"

He starts to tell me about telephone pole people and struts back and forth on the porch like a windup toy, walking stiff-legged now, his arms held tightly at his sides.

"Telephone pole people, Lillun, the ones who just go trudgin' along, walkin' on tracks set down by somebody they don't even know."

I jump up and start to mimic him. We walk stiffly, looking down at our feet.

"They walk where somebody say walk, they dance in a special room somebody say dance, they eat in only one place, 'round a table, they sleep in only one special place, a reg'lar bed in a room that's just got a bed in it! They insides is all divided up, just like that. They got no joy spillin' out, Lillun, no love.

"Joy, Lillun!" he shouts, flinging his arms wide and turning in circles, faster and faster. Leaping from the porch he reaches for me. We join hands, we whirl on the hard-packed dirt until we are dizzy, the chickens scatter, we fall in a tangle, laughing.

Puppie sees us as he comes down the road. He brings

a bag of potatoes for our supper. He calls to us, and we run to meet him. We make a game with the potatoes. We throw them to one another, we toss them in the air, we run, we dance, we chase, we tumble. The sun smiles down at us.

Bigun races me back to the porch. He wins because he jumps over the railing. Puppie and I take turns throwing the potatoes to him until Bigun has caught them all and lined them up on top of the railing. "I'm the king of the potatoes!" he calls, trying to balance one on his head as a crown.

We toss the potatoes into a bushel basket. One of the chickens is on the porch, and Bigun picks it up, puts it on top of the potatoes. "Is you the goose that laid all them golden eggs?" he asks solemnly, leaning over to the chicken.

Puppie has said, "Iffen you got chickens, you got eggs, and iffen you got eggs, you got chickens. And that's all anybody can need. That and potatoes and each other."

After Sally leaves, I walk around her apartment. It is mine, too, I remind myself, but it seems to belong to Sally, like her lovely dresses, like her stylish friends. I walk around, touching the smooth surfaces, and I say to myself over and over, *I will remember—when the time comes, long from now, when I want to remember this year, I will remember this place.*

Will my memory be sharp and neat, I wonder, like this room, like Sally, or will the edges be blurred and soft, like my memories of my times with my father?

And what part of this time will I remember? What will pull me back here someday, to see again in my mind these bright paintings on the walls and remember the feel of the polished wood and smooth bright expensive fabrics?

Sally has given me a prescription to fill at the neighborhood drugstore, a prescription for her headaches. She says I may charge it. She has opened charge accounts in all the nice stores.

When I walk into the drugstore, a girl my age is there looking at different kinds of shampoo, different kinds of perfume. She is a very pretty girl. She is dressed in pale green slacks and a blouse to match. Sally has taught me how to tell by looking at something whether it is good or not, expensive or not, by the material, the cut, the way the collar lies, even the buttons, and I know that the girl in green is wearing expensive clothes. I notice too her expensive shoes and the watch on her wrist. The man behind the counter takes the prescription blank and asks me to wait a moment.

I know the girl is looking at me, deciding about my clothes, wondering about me. When the pharmacist gives me the bottle of pills, he asks if I want to charge it. I nod, and he says, "And what is your father's name, young lady?"

The girl in green listens, and I say, "I don't have a father. My mother's name is Mrs. Sally Goodhall. We live in the Stanton Arms."

Maybe now the girl will walk over to me, maybe she will say, "I heard you say that you live in the Stanton Arms. Well, I do too. Isn't that a coincidence?" and I will take her up to meet Sally, I will show her our beautiful apartment, I will show her my perfect blue and silver room.

But the girl says nothing, and I put the bottle of pills in my pocket and leave the drugstore.

On my way to see Maisie, I stop at a store next door and buy a plant, a little green plant. I give it to her when I get there, and she ties a ribbon around the pot and puts it in the window.

"A present somebody give you must be the quickest way to say scat to the Sad One," she chuckles.

I tell her about the time of the dandelion.

There is a yard in the back of where we live, a yard we share with the other families who live in our building. There is no garden, there are no bushes, no swings or slides. My father calls it the scrubyard. The dandelions grow, and one day Bigun shows me one and explains to me that it is a magic flower, beautiful and strange. But he says nobody appreciates dandelions because they grow in scrubyards like ours. He builds a little fence of chicken wire around one of the dandelions, and we watch it each day, watch it turn into another kind of flower, a soft round gray flower with tiny seeds that have wings, Bigun tells me, and the wind carries them away to make new dandelions.

"Weeds, plain-growin' everyday weeds, plain-growin' everyday vegetables, they's sometimes so beau-

tiful and so chuck full of God it can make you cry. Some folks can't like nothin' plain, that's the truth, Lillun. Some folks, they turn up they noses at things that come easy and every day. They pamper flowers, grow some special, twist 'em to they own ways, has 'em in glass houses, then they like 'em!

"Some folks hate pigeons or sparrows 'cause they think that kind of bird must be a nothin' kind, just 'cause they see them birds every which way they look. They get theirselves expensive birds in cages, them birds die if they get outside.

"Some folks don't pay no mind to reglar cats, alley cats. They got theirselves cats so special they ain' allowed noplace. They taken the claws offen them cats, so they can't climb no trees, they can't get they rightful food, they can't fight, they can't nothin', but they mighty purty iffen you like just purty. And who's to say what's purty, what's beautiful, what's not, Lillun? Who's to be the decider?"

"I don't know," I say.

"Each can be his own decider, each can be his own chooser," says Bigun.

"We had a cat," I say now to Maisie, and I tell her about the time Sally had given me a cat. It is a very special, very exotic breed. From the beginning the cat and I do not like each other. He does not look at me, maybe because he feels I am not good enough, maybe too because he can sense that I am self-conscious and uneasy with him. Cats are not allowed in our new apartment, no one at Stanton Arms may have a pet, and I am glad when Sally gives the cat to one of her friends. I am sure the cat is glad too, because the friend is like the cat, elegant and smart, beautiful and cold.

And Maisie tells me more about the Sad One. "If you go outside too much by your lone, he's like to spy you. He can tell maybe here is somebody lonesome, somebody to move in with. So if you got to go to market, you go real fast, you act like you got to hurry, you act like somebody waitin' for you at home. Sometimes when you come back, maybe the Sad One is waitin' outside to see how you act, see if you got anybody for real or not. So you pretend like you glad to get back, you wave like somebody's inside waitin' for you, you knock at your own door and you say, 'Hey, open up! What's that you doin' in there, sleepin'?' " She laughs. Her laugh is like my father's laugh, her eyes are like my father's eyes.

The next day while I walk I see some girls standing and talking, not far from the school. They are about my age. I feel shy, but I walk proud and pretending I am full of secrets, the way Sally has told me. I can tell they are looking at me, I can feel that they are talking about me now, maybe sizing me up, deciding things about me from the way I look. I am glad I am wearing my red silk scarf today. I wonder if I should look at them, I wonder if I should smile. Sally would know.

The girls wear bright colors, they are like birds together, I feel they are about to fly. I start to walk over to them, but they run into one of the apartment buildings.

Maybe they live in there. Then they will go to the same school I will go to. They will be my friends, if I play my cards right. We will laugh together and have secrets. When, I wonder, when?

I pretend to be looking for someone to come, for something to happen. I look at my watch, I walk up and down, and I wait. Maybe they are looking at me from one of the windows. Maybe they are saying, *I wonder who she is, I wonder what she's like.* Maybe they are saying, *Wouldn't it be fine if she could be our friend?*

I walk, I hold my head tall, just as Sally has told me. Sally would be proud of me.

After a while, I go home. But now each day I walk here, waiting for them to see me, waiting for them to be my friends.

Maisie tells me stories her grandy has told her, and I tell her stories I have read long ago in beautiful books with shiny pages, stories about princes and princesses, and I tell her about the lovely bright pictures.

"I never did see a book like that," says Maisie. "All my pictures are settin' in my head."

I tell Maisie I will find her such a book and bring it to her that afternoon.

When Sally leaves after lunch I do the dishes, and then I look for a book to take to Maisie. I take down a book from my shelves. Sally has shown me to wash my hands before I open one, I have learned never to turn the pages down, never to leave a book on its face, never to carry a book around or fall asleep reading.

On the flyleaf in Sally's hand is written: *Dear Sara, You are too young to appreciate this book now. Maybe when you are old enough, we will be together.*

There is a date, it is a Christmas a long time ago, it must be when I was still with my father. I leaf through the book. The leaves are turned down, there are finger marks and jam marks on its pages, and there are drawings on some of them, childish scrawls that I recognize are mine. This is a book from the old days.

I leaf through it, not reading it but looking for clues, looking for bits and pieces of the time with my father, and I turn a page to find an old faded snapshot. I lift it out to look at it. Three people in a funny pose: a man holding out his arm at a right angle; a boy fits just under it. The boy holds his arm out the same way, and a little girl fits just under his arm. The little girl holds a doll out at a right angle. I look closely. Is the little girl me? I can see it is me, but what is there of me that is like, what is there of me that is left?

I turn the picture over. On the back is written: *Puppie, Bigun, Lillun, Weeun.* I shut my eyes and say the names over and over, and they are familiar as an old melody. I look at the picture again. My father and me. My father, Puppie. Bigun and me, Lillun. My doll, Weeun. I look into the face of Bigun laughing down at the little me, the long-ago me, Puppie, Bigun, Puppie, Bigun slip back again into my memory, onto my tongue, and I say them to myself, *Puppie, Bigun, Puppie, Bigun.*

I carry the book and the snapshot into the living room. I stare at the snapshot, and the face of the little girl that was me smiles at me. Maybe she tries to make a friend of me. I try to smile back.

I look out of the window to the apartment building across the street for a long time, trying to remember, and suddenly I catch myself smiling at one of the stories of the Sad One Maisie has told me this day: "You don't never set watchin' out a window, just plain watchin'. Even if you feel like lookin' out, not even lonesome, just tastin' the day, you got to pretend you busy, fussin' with things, too busy to be lonesome none. If the Sad One come passin' by and see you just settin', lookin', he guess you must be lonesome, and he sneak in sure."

I remember the time of the soap bubbles.

We are sitting at a city window with a dishpan of water and a cake of yellow soap. With an eggbeater we churn the water until we have soapsuds. With a new bubble pipe we take turns blowing bubbles through the window. Bigun lets me have the most turns, and after a while he runs outside and whirls in the soap bubbles I make, catching them, making a game of pretending sorrow when they break before he can kiss them. He dances with joy in the soap bubbles and I laugh so hard I can't blow any more bubbles, and he pretends to cry, throwing himself on the sidewalk.

I go into the kitchen, the spotless neat kitchen, with the new pots and pans hanging brightly on the wall, lined up according to size.

I open each drawer, searching for an eggbeater. There are many things in the shiny drawers, all in order. Everything in the kitchen is in perfect rows or stacks. There is no eggbeater.

It is a long-ago day. I am standing outside on the sidewalk in the bright cold sunlight. Scattered around me are my most prized possessions, my stuffed animals, my games, my dolls. I have found them all in the garbage cans set out on the street. I am gathering them together to take them back inside.

Puppie would never have thrown them away, or Bigun. But Sally has come, after a long long time Sally is here, and there is no easiness now. She runs out of the building, down its broken steps. She is wearing blue jeans and a bright red shirt, a bandanna tied around her head. "Trash, all trash!" she cries. "Will you be a garbage sorter all your life?"

I think I start to cry, and Sally shakes me. "You're leaving this junk, you're leaving this place, you're leaving them, you're starting fresh, starting from scratch, coming with me."

She is angry, but suddenly, mysteriously, she kneels down beside me, her lovely earnest face close to mine. "You'll see, Sara, you'll see. If you take stuff with you, it'll be like little pieces of this place pulling at you, keeping you from thinking straight, keeping you from thinking ahead."

There in the trash is a doll with a sewed-on mouth and button eyes, dirty because I kept trying to feed it. My father would tuck me in with it and tell us a story. "Weeun, you like to be fallin' asleep 'fore I tell Lillun the best part of the story," he'd scold the doll.

Of course I don't know then that Sally is taking me away from my father, I only know that she is throwing away Weeun. I keep crying, I remember now that I keep crying, and Sally pulls me into the building and up to our flat. She puts me in a bubble bath. I have never had a bubble bath before. She piles soapsuds on my hair and shows me in the mirror, and I stop crying and I laugh.

While I am still in the tub, she shoves my old dress into one of the boxes. There is a room filled with boxes. These are the things that Sally will be giving to charity. The toys that sit in the trash cans outside are not good enough even to give away.

She shows me my new dress, packed in tissue paper. It is a present from Sally. She helps me with the buttons. I feel strange when I put it on, it does not seem to like me, and I wonder if I will ever be able to sit down in this dress, or run, or play. But it is a fine dress, very beautiful, very expensive. I know this because Sally tells me so.

I still do not know that I am leaving my father, there is never one time that it is goodbye. I know I see him a time or two after Sally takes me away. I remember he takes me to a park, we have a picnic. He pushes me on a swing, and together we slide down a steep slide, laughing. Suddenly he holds me close, his laughter seems to rock his body, but his face is next to mine and it is wet. I taste his tears as if they were my own.

"They ain' no way, Lillun," he whispers. "I tried so bad, but they ain' no way."

Sally keeps me busy going new places, learning new things. She teaches me to like pretty dresses, different food, she shows me new table manners and tells me how to talk. She shows me how to be neat, with myself, with my room, my clothes, my books, my desk, my drawers. Even my days are neat and tidy, with special times for special things. I have a toothbrush and a cup of my own in the bathroom, I wear pajamas now, and I have a bag for laundry on the back of my closet door.

At first every drawer in my room has a label, and Sally has taped labels on the hooks in my closet, for there is a place for everything. Even my bright new toys and dolls have a place. Sally says that if things around you are in order, your head is in order. She says that you must throw away everything you don't need, just the way you must throw away everything in your head that you don't need, old memories, old times. Only keep in your head the things that will get you somewhere.

I do not know yet that I have left my father, but there is a night when I cry for my doll Weeun. Maybe I cry for my father too, and for the nights he would tell Weeun and me stories. I remember now that Sally comes into my room and turns on the light and sees me crying. She is angry, I have always been able to feel when Sally is angry, even when she smiles, but she comes to sit on my bed and she says, "Tomorrow you and I are going into the city and we will buy a new doll. What do you think of that?" And Sally tells me about the new doll and the clothes we will buy for it, a different dress for every day, and perhaps now I stop crying, for Sally laughs and leaves the room.

I do not now remember that new doll, or the dresses for it, but I know it would have been beautiful, the way Sally said.

That afternoon I take the book over to Maisie and read her some of the stories and show her the pictures. I tell her about the soap bubbles, and we find an eggbeater in a jumbled kitchen drawer. Tomorrow I will buy a bubble pipe and we will blow the bubbles out of her window when I come again.

"That Bigun you tell about," says Maisie. "What is he, a friend?"

I nod.

Sally often asks me, am I making friends? I lie and say yes; after all, the girls I have seen are nearly my friends. If I work it right, I will meet them soon.

Maisie tells me more about the Sad One.

"If you set by yourself much, if you have a special place, a comfy, easy place, then you never dast leave it empty. You got to leave somethin' there, each time. Otherwise he sneak in, set in your place real quiet, and then he get you when you come back."

I have never noticed before that Maisie always

leaves her knitting on her rocking chair when she gets up, but now I notice.

"You're just making it up, about the Sad One," I say.

She stops rocking and looks at me with her soft brown eyes. "No funnin' 'bout him," she says. Then she leans back and laughs. She laughs and laughs.

"There ain' nobody would make him up anyways," she says, laughing still and wiping her eyes on her sleeve. "If you was to make up a body, you'd make up somebody else, somebody good. My grandy told me, if you make up a body and think on it a long time, that body starts comin' real."

Later she tells me more, because I ask, "How do you know about the Sad One? How did you find out?"

"My grandy, she told me. She told me for sure a lot of things they just findin' out now. Maybe her grandy told her." She looks at me sideways. "He ain' never gonna get in here to stay, you bet. I can fox him good." She shakes her head and frowns. "It ain' easy. You got to keep watchful. He knows when the time is come, he knows when he can slip in. And when he come, he come to stay. Ain' no use to try and fool him then. He don't pay no mind to them jokes and tricks when he come to stay."

Maisie asks if I will walk outside with her. "That way the Sad One can see I have somebody and he'll go look for some other old lady to stay with, quit followin' me. He been watchin' me lately. I could taste him yesterday in my place."

On a nice cool day, I take Maisie for a walk. She wants to see where I live, just from the outside. "That way I can picture," she says. "I think it must be a place like Corrie live in. If I see where you live, I see where Corrie's at."

We walk slowly. She says she likes to feel the sun coming in warm and the breeze coming in cool. We are just turning the corner to my apartment building when I see the girls ahead. They are the same girls I have waited for, the ones who are going to be my new friends.

I quickly look over at Maisie. Her old sweater is bunched together on one side, the sleeves rolled up, her slip shows, her stockings sag, she wears a child's knit cap on her head, a blue cap with a tassel on top.

The girls will think she is my mother, maybe; maybe my grandmother. Later they will see my mother, my beautiful stylish mother. But that will be later, maybe too late.

I take Maisie by the arm and pull her back. "I forgot something," I say. "I have to go home. You have to go back alone. We'll see where I live tomorrow."

I turn to leave her.

"The Sad One see you leavin', he gonna come for me," she says.

I am angry. The girls will be gone. "I have to hurry," I say. I turn away from Maisie and I run, run to my apartment building, run toward the girls who wait bright as birds.

I run, my heart pounds, I pass them as they cross to the other side of the street. One of them waves, I wave back. I run into my building. There, they can see where I live, they can see it is a pretty place, an expensive place. They will want to be my friends. I am sure they have not seen Maisie.

Later, long after the girls have gone out of sight, I go to find Maisie. I look for the cap with the tassel.

She walks alone, her head down, on the shady side of the street. I catch up to her, take her arm, tell her a lie, and after a while it is all right. But it is not until later, when I tell her we will go to where I live and have chocolate there with my mother, that she looks at me again. While I say it, I believe it. But I know somehow I will have to be more careful after this. I can make up some excuse for not going with Maisie outside.

But after I have left her, I remember the time of the leaves.

I am collecting fall leaves, gold, rust, red. Am I in a park? Am I in a stranger's yard? I do not remember now, but it is strange to me and very beautiful, and mysterious to find that the leaves that were green are transformed.

I am going to put the leaves I gather in a scrapbook. Of course I am older now. I know now that leaves dry and crack and turn to dust like old memories, but this is long ago, and I think I can keep my glowing leaves forever and ever.

I lean down to pick up a perfect scarlet leaf, not an edge of it torn, not a part of it spoiled, an elegant and proud leaf. I will put it on the first page of the scrapbook, it will be the king of the leaves.

A drab little leaf, curled at the edges, clings wetly to it, and I brush it off. I look at it, lying dismal and forlorn at my feet.

Suddenly I know that for a moment, just for a moment, the dull little leaf thought that it had been chosen. I bend down to pick it up, I exclaim over it, marveling. "You are the best," I whisper to it, "the best. You are my favorite and I love you."

I will put it on the first page of my scrapbook. I will draw beautiful designs around it. It will be the main one, the king, the queen, the lovely one, the treasured one.

I collect leaves all afternoon, and that night I spread them out on the radiator cover to sort, and I put up a card table. I have my crayons for the designs, and my paste, and wax paper, and Scotch tape, and the scrapbook.

Sally walks through the room, straightening pillows, pictures, books, and ashtrays. With Sally, things must be right and in order. She admires the leaves, bright and beautiful on the radiator cover, she praises my project, she hums. Sally is always humming.

Then she picks up the dull little leaf. "Here, this one is here by mistake," she says. "It doesn't belong." And she crushes it in her hand and tosses it into the wastebasket.

I keep working at my designs. I decide to put a beautiful gold leaf and the perfect scarlet one on the front page, to be king and queen of the leaves, and I draw for them jeweled crowns.

I do not think of the little leaf again until today.

When I come in, Sally is pacing, her short pink dress swinging gracefully as she moves. "Someone tried to call you," she says, straightening a picture on the wall. "Long distance, person to person. I only heard the operator."

I frown, puzzled. "Who was it?" I ask.

She glances at me sharply. "You don't know anyone it could be, do you?"

I shake my head.

She starts to move lightly around the room. "It's him," she says. "It's got to be him."

My heart pounds.

"What does he mean, calling you? How does he dare? I've warned him." She never says my father's name because she hates him so much.

"Is it my father?" I ask. "Where is he? When can I see him again?"

"When can you see him again? Your father?" shouts Sally, ablaze with anger. "Your father is an *insect!* Lets everyone walk all over him. Lets people push him around. He's a *nothing*. A zero! I took you away from all that. If it hadn't been for me, if I hadn't worked it out, you'd still be stuck with him," she says, stubbing out her cigarette.

Everything Sally does is pretty to watch. Smoking, or setting down a coffee cup, or pushing her plate away, it is all graceful, it is all lovely. I try to watch so I can be like Sally.

It's been a cold time, cold and gray, and when I see something glinting in the sun, I touch it and my fingers stick, frozen. My father comes to help me, and he says, "Just because something shines, that don't mean it's warm, Lillun."

"Why can't I see him?" I ask, and I tremble.

"Listen, Sara," she says, and her anger darts around the room like electricity, like a spotlight. Everything is a brighter color now, the furniture, the walls, the carpeting, the paintings, all take on the glow of Sally's fury.

"I've made sacrifices for you! I've worked hard. And it's not going to be for nothing! Look, look around you! I've given you the best, haven't I? The best!"

The room is dazzling bright now, burning in Sally's rage and contempt.

"And what has he ever done for you? Nothing, Sara, nothing! Oh, he stirred himself once when you were a baby. He moved to the big city with you, he tried it. Became a big businessman! Got himself a little cart that he pushed up and down the street. Selling junk!"

Sally's arms hug each other, hug her, to keep her from flying into tiny sharp fragments of hate. "Where would you be if I'd left you with him?"

She reaches for a cigarette, holds her jeweled lighter to it, the flame leaps up to mirror her anger. Sally inhales new strength, new joy, in her fury.

"He's a lazy good-for-nothing! He thinks life is one long funny joke, one long lazy ride. A ride on a merry-go-round!"

Now she talks quickly, animatedly. She is performing now—not for me, but for her audience, the invisible audience that watches Sally always. I am an extra on the stage around whom her electric performance revolves.

"A merry-go-round," she repeats, "a merry-go-round in slow motion. Come on, hop aboard, have a free ride! Have a good time! Let someone else do the worrying! Let someone else do the working!" She turns to me, blazing and beautiful.

"He thinks that because time is going by, he's the one that's moving. There he is, dazed by the lights, hypnotized by the music, circling round and round, seeing the same things over and over, day after day, year after year!"

She inhales deeply, and when she exhales the smoke curls up and touches her face shyly, tentatively. She turns from it with contempt.

"Oh, maybe on a big day, a really big day, he stretches, he walks around the merry-go-round, he says to himself, 'I am really going to go somewhere today, I am really going to get someplace for sure. This horse I have been riding is too slow, I will find me another horse.' " She stubs her cigarette out with venom. "Stumbling through never-never land, cloud number nine next stop! Cloud number nine!"

Bigun and I are sitting in the scrubyard. We are counting clouds. He points to one and shows me that now it looks like a frog, now a bent little man with a cap, now a car with a mattress on top.

"What do you see, Lillun?" he asks me, and I say, "A cloud."

He laughs, his warm delight spilling over the day. "A cloud, Lillun! That is got to be the most and the best that any soul can see in it. A cloud, a plain simple cloud, is a right wonderful thing. Sure puts to shame a car with a mattress on top."

Now his soft loving laughter comes through the years to fill me, to gentle me, and I ask Sally, "Who is Bigun?"

Sally looks sharply across the table at me. "Bigun!" She spits the name out like a seed.

"Who is he?" I ask again.

Sally stands up, starts to pace around the room. "A lazy, slow, stupid ox, that's who he is. We have nothing to do with him. Bigun," she says scornfully, half laughing. "Even his name is nothing, a name for a baby, a name for a clown."

She walks to the window, turns to look at me and I can see that she has decided to tell me about Bigun.

"Your father's son," she says, mockingly. "Not mine," she adds quickly, lifting her lovely head proudly. "He's got nothing to do with you and me. Bigun was born before I ever came into the picture. They're two of a kind, Bigun and your father. Two for a nickel, dime a dozen."

Her anger has spent itself, she leans earnestly toward me. "You have to understand, Sara. I left you there, but I didn't abandon you. I left you so I could make a place for us, for you and me. It wasn't easy, I had a hard time, but I made it, I'm making it, and I've taught you what life is really about, haven't I, baby?"

Her eyes shine, but not, I know, with tears. Sally never cries.

"You've outgrown them, Sara, just the way you've outgrown your old clothes."

"But he's my father," I say. "Bigun is my brother."

"You're not cut from the same cloth! You can't pull them up. They don't know where up is. They can only pull you down. Look, Sara, I wasn't much more than a kid like you when I was married. I hadn't had any advantages, I hadn't ever had anything or anyone to judge by. He had a big smile, and that was enough." Her lovely lip curls delicately, scornfully. "Well, it's not enough! He's a happy-go-lucky, good-for-nothing failure! And now that I'm really making it without him, really getting somewhere, I suppose he's got to see what's in it for him, what he can get out of it. Well, there's nothing for him, not now, not ever. Not from me, not from you."

Her long slender fingers close over mine. "I'll tell you what we'll do tomorrow. We'll go out for lunch, just the two of us, just you and me, somewhere special, and then we'll go shopping, get some new clothes. How would you like that?"

"Fine," I say.

She paces around the living room, lightly, gracefully. "If he calls again, you hang up, do you hear? Promise?"

I nod.

I tell Maisie the next afternoon that my father has tried to call me. "After all these years, suddenly he calls me," I say, puzzled.

She smiles and says, "I knowed it, I knowed he'd call." She tells me I've been sending out little messages in the air. "Little hummingbird signals. They buzz at his windows, they buzz at his dreams, they get through to him in his sleep and he wakes up saying, 'Now, what put me in mind of Sara?' "

"He calls me Lillun," I tell her.

"Then I call you Lillun too," she says. She nods and rocks. "Them signals, they work. I think on my Corrie, so purty soon she be thinkin' on me. That's how I know for sure she be comin'."

"Why don't you call her?" I ask. "We could go to a pay telephone. I could help you."

She shakes her head. "I dunno where my Corrie's

at. They moved from they old place, I moved from my old place. Ain' no way one of us can figure out where the other one is. So with Corrie and me it just plain gotta be them thinkin' messages I told you 'bout. If it works okay with your daddy, it'll work with my Corrie."

"But he knew where to call me," I say. "He knew my number."

Maisie shakes her head. "My grandy never knew no telephone, she never knew no numbers. She told me this other way. See," she explains, "I send Corrie this here house each day, send her just what it look like. Each mornin', 'fore I even step my feet on the floor, and each night 'fore I drift off, I send her the whole thing, so she can find it easy."

She rocks and smiles. "She be comin', you wait and see. She gonna find me here sure."

When I let myself into our apartment, I see a note from Sally. She will be late coming in, she has left supper for me. I go into my room, my room of blue and ice, and look for something to read.

The telephone rings. The operator says, "Miss

Sara Goodhall, please, long distance calling. Is this Miss Goodhall?"

"Yes," I answer, and in a moment I hear coins dropping, and then I hear his voice.

"Lillun? That you, Lillun? Lillun, it's your Puppie. This bein' away, bein' separate, I feel like somethin's cut clean outta me, cut clean outta my insides, all this time. Say how it's gone for you, Lillun, say how it's gone for you."

I hold the telephone against my ear, and his voice comes into me like warm syrup and makes me gentle and drowsy with love. I had forgotten, but not forgotten, the sound of him.

I think I have said, "Puppie, I love you," but maybe I have said nothing, for in a moment I hear him say, "Lillun, you there, Lillun?" and I long for his voice to cover me, smother me, carry me to him in its liquid softness, and I press the telephone closer to my ear.

"You there?" he asks again, and I whisper, "Yes."

"Lillun, I'm hungrin' to set my eyes on you, I'm so thirstin' for your voice, I'm lonesome for us playin' together like always. Say me somethin', Lillun, so I got somethin' to hold to me."

My throat is so filled with words, with love, with regrets, that I cannot speak. I feel I can lean back in my father's love, fold myself into it, float forever on the tenderness of it.

"Sally, she there?" asks my father. I shake my head, and then when he asks again, I say, "No, Sally's not here."

"I'm achin' for you, Lillun," he says, and now I feel my tears, running down into my mouth.

"You and me, Lillun, you and me together again," says my father, and when I am quiet his voice quickens. "Them bad times, they's over, the good times is comin'. Listen, Lillun, you got to figure what's most, what's truest in your life, what's right and what's no nevermind. Sally, she can give you purty clothes and party times, she can show you good times and city ways. She can fix it so you plenty comfy on the outside. But the inside, the inside is where you gonna be growin', Lillun. Inside your skin is where you gotta be your whole life."

I am quiet, and he talks on. "Them good days,

you ain' forgot?" he asks, and I shake my head. Of course he cannot see that I remember, he cannot see my silent loving tears.

There is silence now, and I try to see him. Perhaps he is quiet as I am quiet, because his throat is filled with memories.

"Bigun ain' here, or he'd say hi," he says in a little while. "Bigun, he been messin' with bad kids, makin' lots of trouble. I don't know what's got into him. You come back, Bigun might maybe straighten out. Iffen you come back, everythin'd be like old times, all us together, you and me and Bigun."

I still cannot speak, and he says, "Lillun, how growed is you now?"

The operator interrupts to say that our time is up.

"Don't go, Lillun, don't go," he cries.

I hold the telephone close, I press it against me, maybe I can bring him back. I say his name over and over until he is drowned in the word, *Puppie, Puppie.*

When I go to bed, I try to find something to put beside me, something of his. I look and I look and finally in a drawer of my bureau I find an old wool scarf, something he sent me long ago, something Sally has not thrown away, because I have hidden it.

I put it on a chair beside me. I am too old to take something to bed with me, but after a while I reach out to it and bring it to me. I think maybe I will go back in the living room to touch the telephone, maybe speak, but I hear Sally come into the apartment with her friends, and anyway I know his voice is cold now and far away.

"Puppie, Puppie," I whisper, again and again, but after a while I know I am only saying his name, and in a few minutes I feel foolish, I start listening to Sally and her friends. They bounce talk back and forth in the room, it is like a complicated ping-pong game with many players, many rules. I wonder if I will ever learn to play any part of that game.

I am not very used to crying, but the warm tears are old and familiar to me now, pleasant and friendly, and I am sorry when they finally stop.

In the morning I find the scarf on the floor, and I put it back in the drawer. I do not say his name today.

The next day Sally tells me that we have an

unlisted telephone number now, and she tells me to write it down and memorize it and give it to my friends. Only my friends will know how to reach me now, the friends I have not yet met.

I walk and walk. I do not even go to Maisie's, I do not even look for friends.

When I get home, I see that Sally is excited about something, excited and nervous. She looks more beautiful than ever, she wears a dress the color of honey, the color of gold. Her long tapered fingers reach up and touch her ears, her earrings. She wears new bright polish on her perfect nails.

"Where do you go all the time?" she asks. "What about the friends you've made, or haven't you made any friends?"

"I have friends," I lie.

"Someone you can stay with for a few days?" she asks.

I shake my head.

She looks at me scornfully, she jumps up, walks to the window, looks out, drums on the windowsill with her narrow lovely lacquered nails.

"I've got to get to New York," she says. "I've got

to go." Now she makes a fist and pounds very slowly, very softly, on the windowsill. "I can't leave you, and I can't take you with me."

I don't know what to say.

"My one chance, my one big chance, and I have to let it pass me by. Everything I've been working for, everything I've been building up to, and I can't go, I can't take advantage of this one wonderful break. I'm trapped! I should have known it was a mistake, bringing you here, trying to raise you by myself."

She turns and looks at me angrily. "I did it for you, to give you a chance. I've put everything in your lap. Why don't you make friends? We've been here long enough, God knows. I can't do everything! I can't do everything all alone!" She storms into her bedroom.

I sit there and look out of the window.

"I know somewhere I can stay," I say finally, walking into her room.

She is banging her bureau drawers open and shut, looking for nothing.

She turns. "You do, Sara? You sure? Somewhere

you can be, just for a couple, three days?" She runs over, throws her arms around me. "You and me, we're a team, honey, aren't we? You can stay with your friends and I'll call up every day."

"There's no telephone," I say. I sense Sally's wandering attention coming to light dangerously on my words.

"She doesn't have a telephone yet," I say.

The danger is past.

"Tell her I sympathize. We waited two weeks for ours. One of the penalties of living in the city. So many irritating complexities."

I walk over to talk to Maisie. She is putting something in the window when I come, a tiny smocked dress that she hangs on a little hanger. She knocks on the window, and I know she has been waiting for me, looking for me.

I ask her if I can stay with her for a few days, and I explain that my mother wants to go away. I say I can sleep on the couch.

She doesn't say anything, and at first I think maybe she hasn't heard me, hasn't understood. Then

I see her wipe her eyes with her sweater sleeve as she turns around.

She smiles at me, but the tears are still in her eyes. She says, "Wait, I got to turn the smile around to the window! If the Sad One be passin' by, he maybe think I cry sad!"

She turns back to the window, she pulls it open and leans out. "See, Sad One, see?" she calls loud. "I got company and I cry happy! That's what that is, Mr. Sad One, happy! Ain' you never seen happy before?"

She brings out a blanket, she rummages around in her closet for extra sheets, she sets out the games we will play together.

"Tell your mother you can stay all week! All summer! All forever!"

Now she wants to go shopping by herself. "It's a secret," she tells me. Cagily, she asks me what size dress my mother wears. She marvels when I tell her size ten. She writes down how tall Sally is and chuckles to herself. "A surprise you can take her tomorrow," she explains. "A beautiful present to a beautiful lady."

The next day I take my things over to Maisie's. She is setting the table. "We ain' playin' make-believe this time! No pretend about this," she says. "You see the new tablecloth?" It is a bright plastic with raised flowers around the edges.

I tell her I am going back to say goodbye to Sally. Maisie gives me a box tied shut with a ribbon. "A surprise for your mama."

I tell her I will be back soon, and then we will have supper, play cards, and this will be the first night I will spend.

As I walk down the steps, I look back. She holds up an eggbeater, nods and smiles.

I walk with the box. I walk to the apartment building where I first saw the girls. Maybe they are looking out, maybe they are saying, *There she goes again.* Maybe they are saying, *I wonder what she is like, I wonder what is in the box, I wonder if we could be friends.*

Then I walk home to say goodbye to Sally.

Sally is wearing a new suit, the palest of pale blues, her matching suitcase closed and waiting.

"What took you so long?" she asks crossly when I

come in. Before I have a chance to answer, she gives me a radiant smile.

"I have a surprise for you, Sara, a real surprise!" She stretches her arms above her head and turns like a dancer. I put the box down on the coffee table and sit on the couch, the new couch with the sleek lines and the soft bright pillows.

"While you were gone, Amelia Carter called," says Sally breathlessly. "I met her at a meeting the other day. She's into the school scene, president of the council, something like that, into everything, knows all the best people."

I watch Sally's moving face, her beautiful changing quicksilver face, and wait for her to tell me.

"Well, Mrs. Carter has a daughter your age. You'll be in the same class in September. I forget her name. But listen, Sara, they've invited you to a swim party this afternoon. At this special club or something. And when I mentioned, just happened to mention, that I was planning to be in New York for a few days, she said you could stay there! Think of that, Sara. You, a house guest at the Carters'!"

I wonder if the Carter girl is one of the girls I have seen.

"What about my friend?" I ask uneasily.

Sally turns on me. "Listen! Listen to me! I don't care about your friend, whoever she is. I care about this chance with the Carters. And that's what you should care about! You've got to learn what's important and what isn't. Who matters and who doesn't. And the Carters matter!" Her eyes are bright and beautiful with anger.

I feel my nails digging into my arms where I am holding them. Sally lifts her head high on her long neck. She looks like a princess in a book. I try to think of something to do, something to say.

"You're going to stay with Mrs. Carter, you're going to meet her daughter's friends, whether you like it or not," says Sally, reaching for a cigarette. She sees the box on the coffee table. I watch as if I am watching a movie. I cannot move, I cannot speak.

She looks at the box and frowns, raising her eyebrows. Then she takes off the ribbon and opens it. She lifts out a dress, a bright flowered dress, a purple dress with beads, a cheap, ugly dress, a dreadful dress, like the ones that Maisie wears.

Sally presses her hands to her mouth. I think maybe she is sneezing, maybe she has a toothache. But she is laughing, she explodes with laughter. She leans against the desk, she holds on to it and sways in her joy.

"I don't believe it," she says finally. "How deliciously horrible!" She wipes her eyes. Sally always carries a fresh handkerchief tucked in her long sleeve. Even if Sally were to cry, she would not forget to use her handkerchief.

"Priceless!" she gasps. "Too terrible to be true!"

The doorbell rings, and I go to answer it. Three girls stand on the threshold, smiling. A lady is with them. They are the girls I have seen, little fledgling Sallys, easy and smooth.

"I've only just now had a chance to tell Sara about your lovely invitation," says Sally. "She'll be ready in a minute. Darling, this is Mrs. Carter. This is my Sara."

Mrs. Carter's daughter is named Rosalie, I don't hear the names of the others.

"We'll take good care of Sara," says Mrs. Carter. "You just forget all about her and have a wonderful time in New York."

Mrs. Carter and the girls look at Sally, they can see how beautiful she is. I feel myself blush, as if they are looking at me. Sally eases them into the living room, they look around, they can see it is a modern room, expensive and tasteful. They all talk lightly, easily, with animation.

I feel my tongue heavy in my mouth, my arms awkward, I cannot find a comfortable expression for my face. Sally excuses herself for a moment, goes into her bedroom, shuts the door. There is a long silence in the living room. They all look over at me, expecting me to say something to show I am worthy of them.

Then I see the dress, the dress that Maisie has bought for Sally. I try to smile. I hold the dress up and I laugh, a laugh like Sally's.

"The funniest old lady gave this to my mother," I say. "Her name's Maisie."

Suddenly I am excited, I am inspired, I am filled with electricity. "Crazy Maisie," I say, laughing and holding up the dress. They smile politely and wait for me to say something more.

"You won't believe this," I say, and I feel their interest start to flow like a warm current between us. Rosalie tilts her head and looks at me. We will be friends, I will show her my blue and silver room.

"Wait till I tell you the stories she tells," I say, "the things she believes!"

"Oh, tell us, Sara," says Mrs. Carter. They all murmur softly in anticipation.

I start to talk quickly. I use my hands to describe things, the way I have seen Sally do. I imitate Maisie's rocking and knitting. I tell about the Sad One, I tell about Corrie. For a moment, I imagine I see out of the corner of my eye someone, something, sitting on the other end of the couch, but it is no one, it is nothing.

I stop to catch my breath. There is a foreign taste in my mouth, there is a lump in my throat, I swallow. For a moment, I think I can run from the room, run to Maisie's, leave them here, leave them all, but of course I never will.

"Crazy Maisie!" I say, holding up the dress again and covering my eyes in mock horror.

I start to laugh. I lean against the soft bright pillows on the couch, laughing, and my new friends laugh with me.

About the Author

FLORENCE PARRY HEIDE is the author of more than a dozen books for young people, including *The Shrinking of Treehorn.* She also writes song lyrics in collaboration with composer Sylvia Van Clief, and together they have published several songs and song books. A native of Pittsburgh, Pennsylvania, and a graduate of the University of California, Ms. Heide lives in Kenosha, Wisconsin, with her husband, attorney Donald Heide. They have three sons and two daughters.